STUDENT SCIENCE ACTIVITIES
FOR GRADES 6-9

STUDENT SCIENCE ACTIVITIES FOR GRADES 6-9

RUDOLPH A. BALDINI

Photography by J. Rutherford Ferrill

PARKER PUBLISHING COMPANY, INC. West Nyack, N. Y.

© 1973 *by*

Parker Publishing Company, Inc.
West Nyack, N. Y.

Library of Congress Cataloging in Publication Data

Baldini, Rudolph A   (date)
    Student science activities for grades 6-9.

    Bibliography: p.
    1. Science--Study and teaching (Elementary)
2. Science--Experiments. 3. Project method in
teaching. I. Title.
[LB1585.B24]   372.3'5'044   72-6755
ISBN 0-13-855817-5

Printed in the United States of America

*This book is dedicated to my wife, Rina,
and children, Kathleen and David.*

HOW TO USE THIS
TEACHING TOOL

The best innovator of methods, techniques and approaches is the experienced teacher in the classroom. He must face the daily challenges the students present. Only a teacher who has taught a particular grade level learns the needs and the capabilities of its students.

This book was written for science teachers searching for science activities that arouse curiosity and stimulate learning in the classroom.

All science teachers know that an activity or an experiment is the best way to reinforce a scientific concept or principle. Science cannot be meaningful to students unless they can actively be involved in the process of science. The practical experiments and science activities in this book have been selected to develop a meaningful supplement to your science program.

General Science is an exploratory course and permits the teacher and students a wide range of science topics. Regardless of the subject chosen, the creative teacher will have to relate the subject matter to the students. Simple do-it-yourself activities will often help give students confidence and develop their interest in science.

Since General Science covers a wide range of subjects, the author has placed topics in physical and biological science together. The fields covered in this book are: Electricity, Time, Flight and Space Travel, Chemistry, The Human Body, Green Plants, and Insects. In each chapter you will find diagrams and photographs that illustrate the activity.

This book concentrates on selected science areas and materials that have been developed in the classroom during years of teaching experiences. All of the presentations can be adapted by the teacher to his own classroom situation. The activities included are for students with a wide range of abilities.

Use the scientific method or problem solving whenever possible. Since these activities are not phrased in such a manner, when you present them to the students, you would rephrase them. An activity should start out as a question the students have to answer. At no point should the teacher tell the students what they will see. It is vital that they draw their own conclusions with proper guidance. Controls in experiments are mentioned. These must be explained to the students, and should be used as often as possible. Above all, the spirit of inquiry should be stimulated. Children, however, cannot be expected to "discover" everything, and some experiments should be planned for them. This book also provides experiments of this type.

I advise the teacher to try an experiment before assigning it to students. All activities will serve as a starting point, and both teacher and students will learn to modify, improve, and expand on some of these experiments with their own ideas. When that happens, your students will indeed feel that science is an interesting and meaningful subject.

Rudolph A. Baldini

CONTENTS

CHAPTER 3 — CHEMISTRY ACTIVITIES

CHAPTER 4 — ACTIVITIES THAT HELP TO EXPLAIN THE HUMAN BODY 109

—1—

ELECTRICITY
ACTIVITIES

The nature of electricity is always fascinating to students because of the marvels it brings to them daily through radio, television, and the other electrical appliances in the home. This chapter demonstrates fundamental working principles of electricity. It also illustrates the production and detection of electricity and the relationship of it to magnetism — thus giving the students a working knowledge of electricity.

STATIC ELECTRICITY DEVICES

How to make a pill bottle electroscope

One of the easiest ways to generate static electricity is by rubbing two unlike materials together. The friction between the objects causes some electrons to move from one material to the other. The object that gains electrons is said to have a negative or minus (—) charge. The object that loses some of its electrons when rubbed is said to have a positive or plus (+) charge.

An instrument which detects static electrical charges is called an electroscope.

It is difficult to produce static electricity on a moist or rainy day. The ideal is a crisp dry one. The teacher should keep this in mind when introducing static electricity activities in order to prevent failures in its production and in the uses of static electricity devices.

Materials:

> Pill bottle, a two-inch piece of Christmas tinsel or any other flexible metal foil, and a 2¼-inch piece of electrical insulated wire.

Procedure:

1. Punch a small hole through the center of the cap of the pill bottle.
2. Strip about ¼ inch of insulation off both ends of the wire.
3. Place the wire through the top of the pill bottle cap to about the center of the wire. Bend the bottom of the bare wire into a "j" shape, and the end into a small loop.
4. Lay the tinsel or foil over the "j" end. Press the two foil leaves gently together. Now place the cap with the wire and the foil on top of the pill bottle.

The class can experiment with various materials by rubbing them together and testing them for an electrical charge. For example, static electricity is produced when a rubber comb or a balloon is rubbed against a piece of wool cloth. Electrons will move from the cloth to the comb or balloon which will then have an excess of electrons, and thus become negatively charged. When students touch the comb to the top of the electroscope, the leaves of the tinsel become charged by the contact and the two leaves have like charges, and will repel each other or fly apart (figure 1-1).

To discharge the electroscope, touch the end of the bare wire loop with your finger. The electrons pass from the wire through your finger, into your body, and onto the ground. The

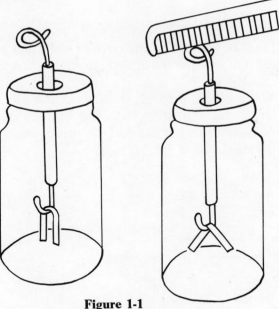

Figure 1-1

foil leaves will then come together again in a neutral position.

Several inquiry experiments and discussions can be developed around the electroscope. Try some of the ideas suggested below:

1. What happens when you shuffle your feet on a rug?
2. Will rubbing two objects of the same material together produce static electricity?
3. Can you feel, see, or hear static electricity?
4. How are devices such as lightning rods and drag chains related to static electricity?

Leyden jar

A Leyden jar is an instrument that stores an electric charge for a short period of time. It was first used in Leyden, The Netherlands, in 1746. These jars were the first form of capacitor. Capacitors are important for radio transmitters and telephones and other electrical devices.

How to make a pill bottle Leyden jar

Materials:

Pill bottle, aluminum foil wrap and a large nail.

Procedure:

1. Line the inside and the outside walls and bottom of the pill bottle with foil up to about ¼ inch below the lip cap (figure 1-2A).

2. Make a hole through the cap of the pill bottle with a large nail and drive it to the bottom. Make sure the point is touching the foil on the bottom of the bottle (figure 1-2B).

Foil inside

Foil outside

Figure 1-2A   **Figure 1-2B**

In order to charge a Leyden jar, rub a balloon three or four times on a piece of wool. The electrons will rub off the wool onto the balloon. This will make the balloon charged. If you touch or place the balloon next to the nail on the Leyden jar, you will hear a crackling sound. This sound is the electrical charge going from the balloon to the jar. Repeat the above procedure for charging the Leyden jar three or four times. This will store electrons and thereby increase the electrical force of your Leyden jar.

How do you know if the Leyden jar is charged? Take the charged jar and bring it close to your ear lobe. If it is charged, you will hear and feel the charge going from the jar to your ear. Another way is by touching it to your electroscope. If it is charged, the foil leaves of the electroscope will repel each other. The electrons may be discharged by touching the metal on top of your electroscope. The foil leaves will then come together again.

You may have students make a list of inquiry experiments about the Leyden jar such as the following:

1. How long will a Leyden jar stay charged?
2. How many different charges can be given off by a Leyden jar onto an electroscope?
3. Can charges be passed on from one Leyden jar to another?

These activities give students a chance to build two simple tools that will not only detect static electricity but store it. Building these instruments can involve every individual in the classroom or be carried out by dividing the class into small groups. In building and using these instruments, students can discover some of the basic laws and concepts of electricity.

HOW TO MAKE AN ELECTRICAL CIRCUITRY CARD

Most students, in studying electrical circuits, cover standard series and parallel circuits that are found in most general science books. These activities are usually done as demonstrations or by a small group of students, rather than on an individual basis, because of the amount and expense of electrical equipment needed. These simple card-circuitry activities that have been used in the classroom will give students, as individuals, the opportunity to apply the principles of circuits by actually designing, building, and testing circuits with the minimum of equipment and expense.

Materials:

1½-volt dry-cell flashlight lamp (with or without a lamp receptacle, electrical wire, 5 × 8 inch cards, and a stapler.

Dry cells and flashlight lamps can be supplied by individual students from household flashlights. The rest of the materials can be supplied easily by the school. The small lamp receptacles are optional.

The parallel circuit is probably the most common type and is used in wiring circuits for the home. Through the use of parallel circuits, it is possible to have a separate switch on each appliance so that they can be turned on or off individually. We may use one socket or outlet at a time, or we may want to use several or all of them at once.

Activities:

The following circuit-card activities will give students an understanding of the parallel circuit in wiring a home circuit by actually wiring a similar circuit onto a card. The circuit diagram may be set up by having the students make their drawings on 5 × 8 inch cards. An example of a circuit problem is to set up two individual wall sockets serving two appliances (e.g., a toaster and an electric iron) as you would find in most kitchens.

The student then lays out the pattern of the wiring to the electrical sockets by drawing it in with a pencil on the circuit card. He then cuts the wire and lays it out on the pattern drawn on the circuit card. The insulation on the wire must be stripped from the ends of the wires whenever the wires are joined together in the circuit in order to make an electrical contact. The wire is then secured onto the card by simply stapling it to the card (figure 1-3).

The student can then test to see if the circuit works by attaching the dry cell at one end of the circuit card and then touching the end of the wires of the socket to a flashlight lamp, with or without a lamp receptacle. If the circuit is working properly, the lamp should light. The other socket can be tested by shifting the flashlight lamp on the left of the diagram and making contact with the wires (figure 1-4).

Caution the students never to plug wires into an electrical outlet. They are often tempted to do so.

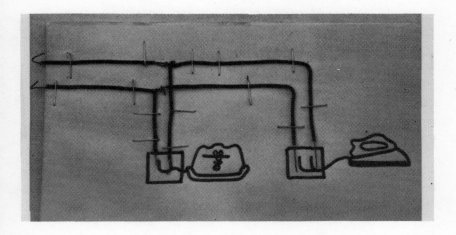

Figure 1-3

Figure 1-4

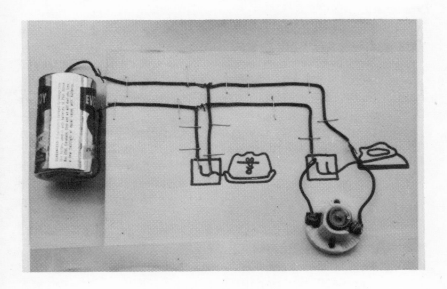

HOW TO MAKE AN ELECTRIC QUIZ BOARD

The electric quiz board may be used as a summary study of electric circuits. The board may also be used to review, summarize and reinforce many concepts, facts and figures in science. This device uses cards that have questions or symbols on one side and answers scrambled on the opposite side. By placing an electrical terminal on a question and another terminal on the correct answer, an electric light bulb goes on.

The electric quiz board described below is only a sample. The size of the board can vary from a mini-quiz board to one as large as the maker desires. This project is suitable as an individual or small group project.

Materials:

> Corrugated cardboard, paper clips, 3 × 5 inch cards, flashlight bulb, insulated electrical wire, 1½-volt dry-cell lamp receptacle

Procedure:

1. Cut a piece of ll × 17 inch corrugated cardboard from a box.
2. Decide how many questions and answers you will have on your board. Place the paper clips on each side of the corrugated cardboard for each question and answer opposite each other.
3. Connect on the back side of the cardboard a piece of wire from the correct "question" clip to the correct "answer" clip. Twist ends of the bare electrical wire around the paper clip. The advantage of using paper clips is that the wire is connected to both the question and answer clips. These clips can easily be moved for changing the questions and answers without disconnecting the wires attached to them.
4. The "tester" may be made by following the photographic illustration (figure 1-5).

Figure 1-5

The lamp receptacle and bulb is tied onto the dry cell. Two wires are connected for the question and answer terminal. Take off the insulation at the ends of the terminals. The tester is used by placing the question terminal and the answer terminal on the correct answer. When both terminals are placed on the correct question and answer sequence, the circuit will be completed and the light will go on (figure 1-6).

The advantage of making the tester separate from the quiz board is that it can be used by many different students. Each quiz board does not need its own tester. It may be passed on from student to student to be used. If the students have any difficulty in the operation of the quiz boards, have them check for defects or faults in the light bulb, dry cell or wire connections.

HOW TO MAKE A MINIATURE
LAMP RECEPTACLE

Many of the electrical experiments and demonstrations cannot be done in the classroom because of the lack of miniature

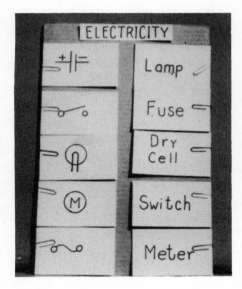

Figure 1-6

lamp receptacles. This is a way of building inexpensive lamp receptacles for student experiments or demonstrations on how a light bulb in a lamp receptacle works.

Materials:

Insulated wire, egg carton or egg crate, flashlight bulb, aluminum foil, masking tape, and a dry cell.

Procedure:

1. Cut two pieces of insulated wire 3½ inches long. Strip from both ends of the wire about 1½ inches of insulation. Now wrap one of the bare wires around the brass end of the flashlight bulb and twist the wire so it will hold fast around the bulb. Bend the rest of the wire attached to the bulb down (figure 1-7A).

2. Cut out a cone from an egg carton. Cut a "×" at the flat top of the cone. Cut both ends of the bottom of the cone about 1/8 of an inch (figure 1-7B).

3. Insert the light bulb with the wire attached into the top of the cone up to the glass of the light bulb. Take

the wire and lay it against the wall of the inside of the cone. Run out the wire through the cut at the wide end of the cone. Tape the wire to the inside wall (figure 1-7C).

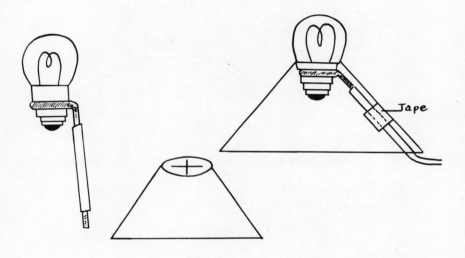

Figure 1-7A   **Figure 1-7B**   **Figure 1-7C**

4. Take one of the 3½-inch wires that you have cut and coil the bare 1½-inch end into a flat coil up to the insulated part of the wire.
5. Cut a 3/4-inch square of aluminum foil and fold and press the foil around the bare coil till about ½-inch square (figure 1-8A).
6. Take the wire with the coil and aluminum and tape it to the bottom of the light bulb in the underside of the cone. The foil with the coil inside should make contact with the metal at the end of the bulb (figure 1-8B).

Touch the two terminal ends of the wire to the poles of a dry cell. The bulb should light. If the bulb does not light, check to see if the foil in the underside is making contact with the

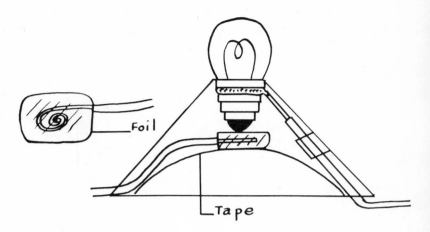

Foil

Ta pe

Figure 1-8A   **Figure 1-8B**

end of the bulb. The foil should be tight against the end of the bulb (figure 1-9).

Figure 1-9

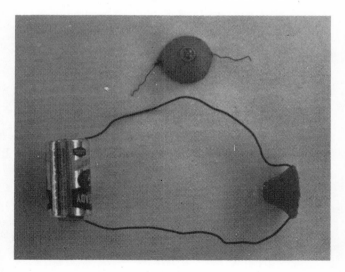

HOW TO MAKE A SILENT OR MERCURY SWITCH

Common switches make or break contact by mechanical movement of a tumbler. The switch makes a distinct "click" each time it is turned on and off as the blades go in and out at the contact point.

A mercury switch is desirable because of its silent operation. A small quantity of liquid mercury is moved against contact points. This type of switch eliminates the wearing out of contact points.

Materials:

> One or two dry cells, insulated electrical wire, test tube, mercury, cork or rubber stopper, miniature lamp receptacle with light bulb.

Procedure:

1. Place a small quantity of mercury into a test tube. Cut two pieces of insulated wire about 12 inches long. Remove about ½ inch of the insulation at the ends of both wires.

2. Place the two pieces of wire opposite each other against the inner wall at the open end of the test tube. Place a cork or a rubber stopper at the end of the test tube and press it tightly against the two wires. About ¼ inch of the bare wires should extend beyond the cork inside the test tube.

3. Connect the wires at the end of the test tube to two dry cells and a lamp receptacle with a flashlight bulb to make a complete circuit (figure 1-10).

4. To complete or close the circuit, tilt the test tube so that the mercury will flow down toward the two bare wires and the cork. The mercury making contact with the wires will complete the circuit and the light will glow. This would be the "on" position in a switch. The "off" position would be set by tilting the test tube

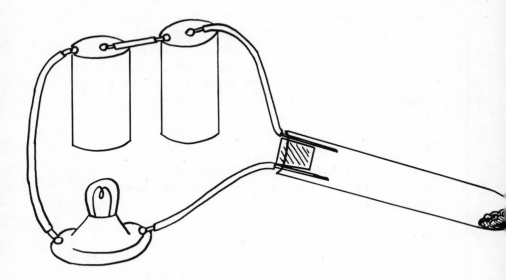

Figure 1-10

so that the mercury would flow down to the closed end of the tube. The circuit would be "open" because of the gap between the bare wires, and the light would go off.

HOW TO MAP OR PLOT A MAGNETIC FIELD WITH A COMPASS

Materials:

Bar magnet or large bolt, magnetic compass.

Procedure:

1. If a large bolt is used, magnetize it by rubbing it against a magnet. Place the bar magnet or bolt on a blank sheet of paper and trace the outline of the magnet. Now place the compass about 1 inch away from the end of the bar magnet or bolt. With your pencil mark a line behind the pointer of the compass needle pointing in the same direction.

2. Move the compass in a counterclockwise circle towards the opposite end of the bar magnet. Keep the
   compass about 1 inch from the magnet. As you move
   in a counterclockwise circle, stop about every ½ inch
   and place another line with your pencil. Move the
   compass until you arrive at the opposite end of the
   magnet. You will see that the compass needle will be
   pointing in the opposite direction at the other end of
   the magnet. Continue moving the compass and plotting with your pencil in a counterclockwise direction
   until you arrive at the point at which you started.
   Try to see if you can join some of the lines to form a
   circular pattern.

The pattern of your plotting should show that the magnetic
lines of force are circular in nature as you moved from one end
of the magnet to the other (figure 1-11).

Figure 1-11

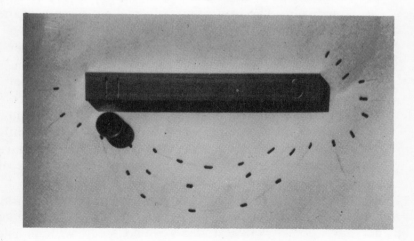

HOW TO MAKE A PERMANENT PATTERN
OF A MAGNET'S FIELD OF FORCE

Magnetic fields are invisible, but it is possible to "see" them and make a permanent picture too by the activity below.

Materials:

Wax paper, Siran Wrap, masking tape, 5 × 8 inch cards, bar magnets, horseshoe magnets, iron filings and a clothes iron.

Procedure:

1. Wrap our bar and horseshoe magnets with a material that clings to them, such as Siran Wrap. The Siran Wrap is used so the iron filings will not stick to the bar magnets, which can be a problem to clean off.
2. Now tape a piece of wax paper to the 3 × 5 inch card with masking tape. Place unlike poles of two bar magnets near each other under the card with the wax paper taped to the top of it.
3. Sprinkle iron filings on the card and tap it gently with your finger. The iron filings will make a picture of the magnets' lines of force.
4. Now connect a clothes iron and let it heat for several minutes. Place the warm iron about three to four inches above the wax paper with the pattern of the magnetic field of force. The clothes iron must be moved constantly in a circular pattern because of the inflammability of the wax paper. The wax paper melts slightly and clings to the iron filings. Let it cool and you have a permanent picture of the magnetic field of force of a magnet (figure 1-12).

Students can create different magnetic field situations and problems. One question that was asked was, "Does a circular magnet have poles?" Give students round steel washers and let them magnetize the washers by rubbing them against a magnet. Have the students carry on the previous activity with

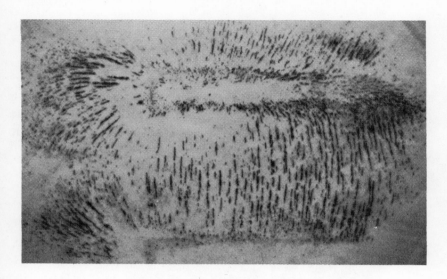

Figure 1-12

wax paper and iron filings and make a picture to see if a circular magnet has magnetic poles.

Permanent patterns can be hung on bulletin boards and studied individually or in a group situation.

HOW TO MAKE A FLOATING COMPASS

This activity will help students who ask, "How do we know that the earth acts as a magnet?"

Materials:

Sewing needle, cork and a dish or bowl.

Procedure:

1. Magnetize a sewing needle by rubbing it with a magnet. Slice about ¼ of an inch from a cork. Stick the needle through the center of this piece.
2. Place the cork slice in the center of a dish or bowl of water. The level of the water must be high enough

so the cork will float freely off the bottom of the dish (figure 1-13).

3. Have the students turn the cork in any direction they desire and let it float freely. They will soon discover that the needle will always point in a north-south direction. Results can be compared by members of the class. The north end of the floating compass needle may be determined with a magnetic compass. This then may show how the earth acts as a magnet.

Figure 1-13

PROJECTS SHOWING THE RELATIONSHIP BETWEEN ELECTRICITY AND MAGNETISM

How an electric current flowing through a wire produces magnetism

The first person to discover that an electric current flowing through a wire produces magnetism was Danish scientist Hans Cristian Oersted in 1819. His first discovery can be duplicated by students in the classroom.

Materials:

Insulated wire, 1 dry cell and a magnetic compass.

Procedure:

1. Measure about 2 feet of wire. Connect one end of the wire to the positive (+) post of the dry cell.
2. Line up the center of the wire over the north-and south-seeking needle of the compass.
3. Connect the other end of the wire to the negative (−) post of the dry cell (figure 1-14A).
4. When the current flows, the compass needle will turn in an east-west direction (figure 1-14B).

Figure 1-14A   **Figure 1-14B**

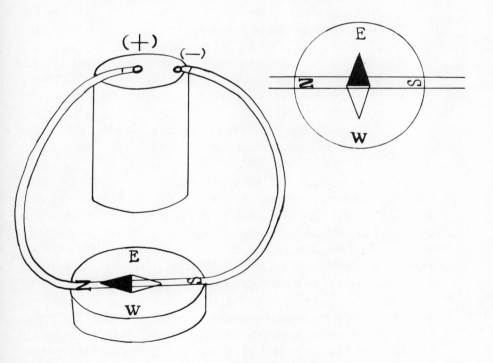

The current flowing through a wire acts like a magnet and attracts the compass needle. The compass needle shows the position of the lines of force in a magnetic field around the wire.

How current flowing through a coil produces magnetic poles

Materials:

Insulated wire, 1 dry cell and a magnetic compass.

Procedure:

1. Measure about a foot of wire for a leader and then continue coiling the wire around a pencil about 18 to 20 times. Leave another foot of terminal wire at the opposite end.
2. Place a magnetic compass at the lower end of the coil of wire.
3. Connect the terminal wire on the upper part of the wire to the negative (−) post of the dry cell. Connect the terminal wire on the lower part of the coil to the positive (+) post.

   Notice that when the current is flowing the north end of the compass needle will be attracted to the end of the coil (figure 1-15).

Disconnect the wire at the negative (−) post of the dry cell. Shift the magnetic compass to the top of the coil. Reconnect the wire to the negative (−) post. You will notice that the south end of the compass needle will be attracted to the end of the coil. This illustrates that a coil of wire with a flow of current running through it will produce opposite magnetic poles at its ends similar to a magnet. Ask your students, "What would happen to the poles at the end of the coil if the current was reversed to flow from the bottom of the coil instead of the top?" The students might write out their theories before experimenting. Reverse the current and repeat the previous experiment. You will find that by reversing the current through the coil, the poles at each end will also be reversed (figure 1-16).

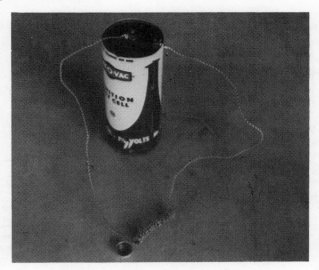

Figure 1-15

How an electromagnet can be made stronger

An electromagnet can be made by coiling wire around a nail and running a current from a dry cell through the wire. The electromagnet remains a magnet only when electricity is flowing through it.

A way of testing to see if electromagnets can be made stronger is by adding coils of wire or by increasing the electrical power. The following problem may be proposed to your students, "Can electromagnets be made stronger by additional coils of wire or electrical power?"

This activity can be done in small groups or as a class demonstration.

Materials:

Three dry cells, insulated wire, two 4½-inch nails, and a box of paper clips.

Procedure:

1. Measure out about 12 inches of wire as a terminal leader. Beyond the 12 inches, wind 20 coils of wire

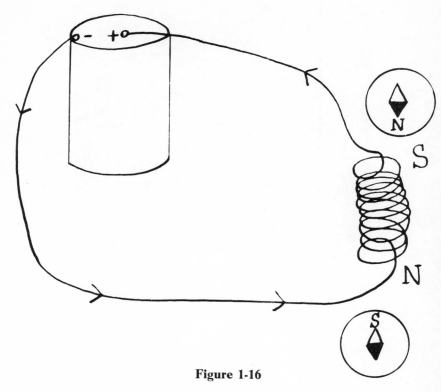

Figure 1-16

around a nail. Keep another 12 inches of wire at the opposite end for a terminal leader (figure 1-17).

2. Repeat step number one on another nail. Coil 40 turns of wire instead of 20.

   The strength of the electromagnet will be measured in terms of how many paper clips it can pick up while the current is flowing through the wire.

   A data sheet can be drawn by the students before attempting to solve the problem (figure 1-18).

3. Connect one terminal ending of the electromagnet to the positive (+) pole of a dry cell and the other terminal ending to the negative (−) post. Dip the head of the nail of the electromagnet into a box of paper clips and lift them out onto the table top. Disconnect one of the post terminal wires. The electromagnet will drop the paper clips. Count the number of paper clips

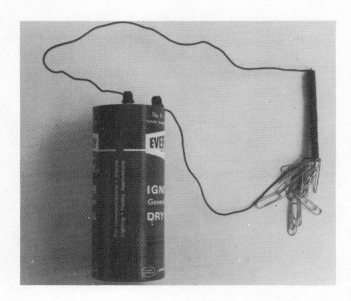

Figure 1-17

No. of trials	1 dry cell 20 coils	40	3 dry cells 20 coils	40
1.				
2.				
3.				
Average				

Figure 1-18

and place the count on the data sheet under trial 1, 1 dry cell and 20 coils. Repeat the experiment two more times and place the counts under trials 2 and 3. Add up the number of paper clips in column one and take an average by dividing by three.

4. Repeat the experiment above with 1 dry cell and use the electromagnet that has 40 coils of wire. Take a count of paper clips and calculate the average as above.

5. The second half of the experiment may be repeated by using three dry cells and 20 coils and then switching to the 40 coils. In each experiment you make three trials to see how many paper clips you can pick up and then add them and take an average. See if the students can write a conclusion based upon their observations and analysis of the data sheet.

Increasing the number of coils or the electrical power will make an electromagnet stronger.

HOW TO MAKE A MINI-TELEGRAPH

The telegraph is based on the principle of the electromagnet. An electromagnet, as we have seen, is a coil of wire wrapped around a magnetic core of metal.

The advantages of building this mini-telegraph are threefold. The first is that it takes a minimum amount of time and materials to build. The second is that because of its simplicity and small size, individual students can build one. (The mini-telegraph is 1 inch wide, 3 3/4 inches long, and 1½ inches high.) The third advantage is that the telegraph sender or key, the device that sends dots and dashes; and the sounder, the device that makes the sounds of the dots and dashes, are both in one unit.

Materials:

Two 1¼-inch penny nails, two pieces of insulated wire (one 17-inch piece and one 11-inch piece), a piece of tin can 3/8 inches wide and 2 inches long, one thumb-tack, one piece of wood 2 inches square and another piece ½ inch high, 3/4 inches wide and 3 3/4 inches long.

Procedure:

1. Cut a piece of wood 3/4 inches wide and 3 3/4 inches long (figure 1-19A).
2. Measure two inches from the edge of the wood and drive the 1¼-inch penny nail into the wood leaving about three quarters of an inch above the wood (figure 1-19B).

Figure 1-19A

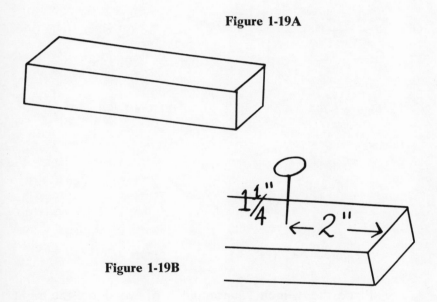

Figure 1-19B

3. Cut a piece of insulated wire 17 inches long.
4. Use 10 inches of the end of the wire as a leader for a connection to one of the poles of a dry cell. Beyond the 10 inches, coil the wire about ten times around from the bottom to the top of the nail. You will end up

with about two inches of wire near the top of the nail (figure 1-20A).

5. Cut another piece of insulated wire about 11 inches long and take off ½ inch of the insulation to the bare wire at each end. Make a J-shaped loop at one end and place it over the left end of the wood. Place a thumbtack at the end of the wood (figure 1-20B).

6. Cut a piece of wood ½ inch square (figure 1-20C).

7. Cut a piece of tin can 3/8 of an inch wide and 2 inches long (figure 1-20D).

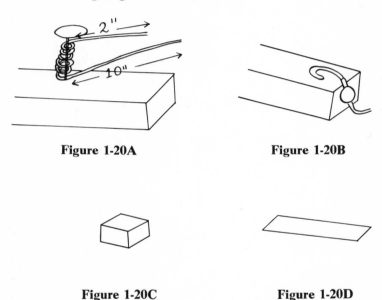

Figure 1-20A   **Figure 1-20B**

Figure 1-20C   **Figure 1-20D**

8. Place the ½-inch square piece of wood on the right side of the 3 3/4-inch long wood. Now place the piece of tin can metal on the top and in the center of the 2-inch square and nail them to the 3 3/4 inches of wood. This makes the three pieces into one unit (figure 1-21).

9. The 2-inch wire near the head of the nail should be bent into a "J" (figure 1-22). This "J" loop of wire placed above the other one serves as the sender and the piece of tin can above the nail serves as the sounder.

Figure 1-21

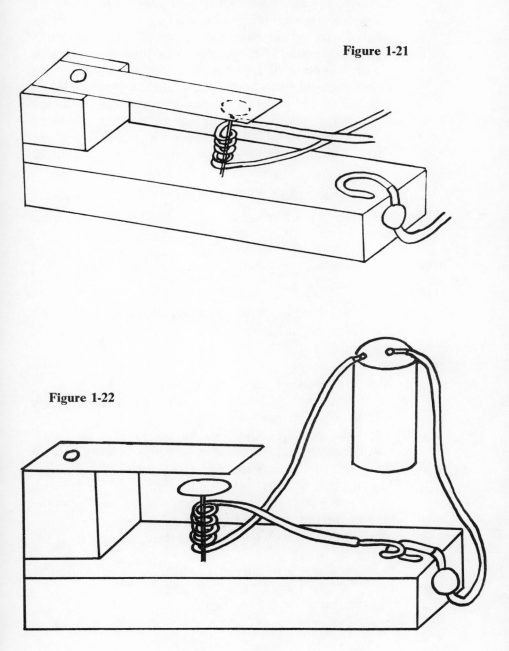

Figure 1-22

Connect the two ends of the lead wire to the poles of a dry cell. By simply pressing down on the piece of insulated wire so the bare end of the "J" makes contact with the second "J" loop, you close the circuit and an electromagnetic nail will attract the piece of tin can above the nail head and make a clicking sound. If the sounder or the piece of tin can is not attracted to the head of the nail, adjust the piece of metal by pressing it closer to or pulling it farther away from the nail head. This in turn is the principle of the telegraph.

HOW TO MAKE ELECTRIC CURRENT DETECTORS

Electric current detectors are devices to detect small amounts of electric current.

Mini-current detector

Materials:

Insulated wire #28, pill bottle, masking tape, corrugated cardboard or wood, thread and a straight pin.

Procedure:

1. Measure out about 5 inches of wire and coil 18 to 25 turns around a pill bottle. The pill bottle is used so you will have a uniform circle.
2. Tape the coils together on the side.
3. Cut the other end of the wire so you will have another lead wire of about 5 inches.
4. Set the coil upright onto a piece of corrugated cardboard or wood about 1½ inches square and tape the bottom of the coil to it (figure 1-23).
5. Tape the two lead wires to the bottom of the cardboard or wood.
6. Now tie a piece of thread to the center of a straight pin.
7. Tie the thread to the top of the coil with the pin hang-

Figure 1-23

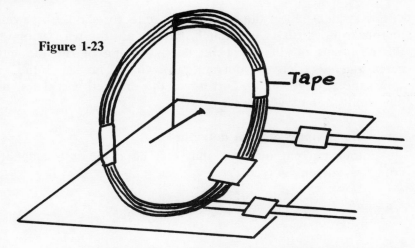

ing level in the center of the coil. Make sure that the ends of the wire are bare before connecting it to a dry cell.

To test the current detector, connect or touch one terminal wire to the dry cell. The other wire of the current detector should be touched at the other end of the dry cell. The wire should be touched and released many times at one end of the dry cell (figure 1-24).

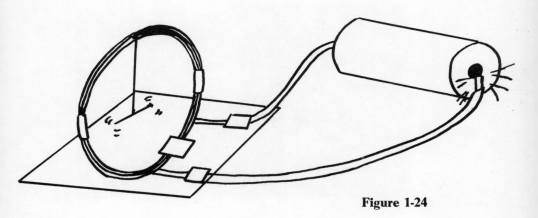

Figure 1-24

The straight pin in the coil of your detector should be swing-ing back and forth as you touch and release the wire at one of the terminals of the dry cell. This swinging of the pin indi-cates that electricity or current is flowing. The faster the pin spins, the stronger is the current. If the dry cell was dead, the pin would not move.

Magnetic compass current detector

Another current detector that can be made is a magnetic compass current detector.

Materials:

Insulated wire, magnetic compass.

Procedure:

Measure about 6 inches of wire and use this as a lead wire for connections. Continue to coil the wire 20 to 25 times around the magnetic compass in a north-south di-rection. Tape the coil to the bottom and top of the com-pass (figure 1-25).

Figure 1-25

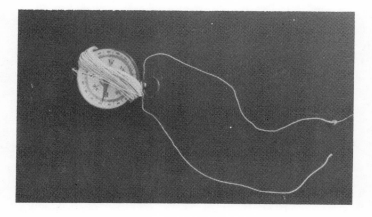

When the wires are connected to a dry cell, the compass needle is deflected by the current producing a magnetic field. The higher the voltage in a dry cell, the faster the needle will be deflected. This is a tool for detecting small amounts of current. Dry cells may be tested to see if they have any life in them (figure 1-26).

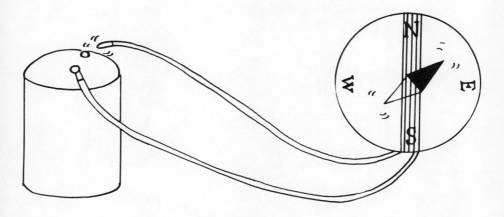

Figure 1-26

HOW TO MAKE A VOLTAIC PILE OR BATTERY

An Italian scientist named Alessandro Volta discovered the electric battery (Voltaic pile). Volta discovered that electrical current can be produced by two different metals such as zinc and copper with salt water or an acid.

Materials:

One square inch of copper and zinc metal, 1 square inch of blotting paper, 1 teaspoon of salt, ½ glass of water. Vinegar can be used in the place of the salt and water.

Procedure:

1. Dissolve one teaspoon of salt to ½ glass of water.
2. Soak the blotting paper in the salt or vinegar solution for a few seconds.

3. Place pieces of zinc and copper metal on opposite sides of the wet blotting paper and squeeze them together. Touch one end of the wire on the current detector to the copper and the other end to the zinc. Touch and lift the wire on the zinc in an off and on manner. If a current is flowing through the wire, the compass needle in the current detector will be deflected or move back and forth (figure 1-27).

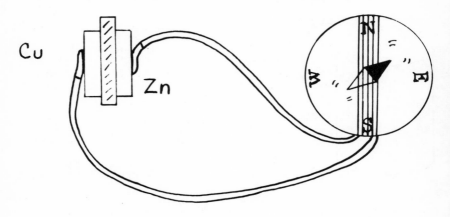

Figure 1-27

A pile or battery can be produced by continuously alternating layers of zinc and copper with moist blotting paper between them. You may make a pile of three or four "sandwiches" and connect the opposite ends to the current detector (figure 1-28).

The greater the pile, the more electric current produced. The greater the current flow, the faster the deflection or movement of the needle on the current detector.

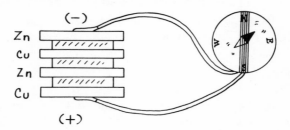

Figure 1-28

—2—

TIME, FLIGHT, AND SPACE
TRAVEL ACTIVITIES

Man has orbited satellites around the earth and has sent manned rockets to the moon. The possibility of exploration of our solar system and the conquest of space has now become a reality.

Today's students have a renewed excitement in the study of the science of flight, rocketry and space travel because of the major breakthrough in space exploration. This excitement and the expansion of space science frontiers have inspired a new interest in the scientific principles of space and rocketry. Because of this new surge of interest by students, the activities and experiments in this chapter place emphasis on stimulating and reinforcing some of the concepts of time, flight and space travel.

47

HOW TO FIND NORTH

This is a simple way to find north during the day without a magnetic compass.

Materials:

> 5 × 8 inch cards, pencils, a wooden matchstick, and a wrist or pocket watch. This activity will have to be done outdoors because you will need the sun.

Procedure:

1. Hold your watch level in sunlight and place a matchstick in the center of the watch face. The match will cast a shadow on the face.
2. Turn the watch until the hour hand coincides with the shadow of the matchstick. North will be halfway between the shadow and 12 o'clock. Since most students do not have watches at this grade level, the following procedure has been used on 5 × 8 inch cards:

    Hand out a blank 5 × 8 inch white card to each student. Have them make a circle with a compass in the center of the paper. By looking at the classroom clock, have them make a copy of it by placing the numbers within the circle from one to 12. Take the students out into direct sunlight and have them place the flat end of their pencil in the center of the card. Turn the card so that the shadow coincides with the hour hand. This is when the teacher with his watch tells the students where the hour hand is during this exercise. Suppose the teacher tells the students that the hour hand is at two. They turn the card until the shadow from the pencil coincides with two. North would be halfway between 12 and two or at one o'clock (figure 2-1).

Figure 2-1

HOW TO FIND EAST AND WEST

Materials:

Broomstick or wooden stake.

Procedure:

1. Drive a stake so that at least three feet of it is above the ground. This should be done during the morning hours in an open area.
2. Mark the spot where the tip of the shadow is resting with a small peg of wood or twig.
3. The two markers will show the directions of east and west. the large stake with the sun behind it will indicate east while the shadow cast by the morning sun and marked by a peg or twig is west (figure 2-2).

Figure 2-2

HOW TO MAKE A SAND JAR CLOCK

Materials:

Two jars of equal size with screw caps on them and some sand.

Procedure:

1. Unscrew the two lids of the jars and cement or paste them together back to back.
2. When the lids are dry, punch a small hole with a nail through the center of the lids.
3. Fill one of the jars three-quarters full of sand. Screw both jars on the cemented lids. For the sand clock to work, turn the jars so the sand will fall into the lower jar.

Have the students mark intervals of time on their sand clocks by looking at the seconds and minutes on a classroom clock (figure 2-3). The amount of time recorded will depend upon the size of the hole in the lids and the size of the jar.

Figure 2-3

HOW TO MAKE A TEST-TUBE SAND CLOCK

Materials:

Two test tubes, two one-hole rubber stoppers (No. 0 or 1), glass tubing 2" long and 3/16th of an inch wide, fine sand, triangle file, and an empty quart-size milk carton.

Procedure:

1. Fill a test tube 3/4 full of sand.
2. Place a one-hole rubber stopper (No. 0 or 1) in the opening of the top of the test tube with the sand.
3. Cut a piece of glass tubing two inches long with a triangular file.
4. Insert the tubing in the top of the one-hole rubber

stopper that was placed into the opening of the test tube with the sand. Leave one inch of the glass tubing above the stopper (figure 2-4).

Figure 2-4

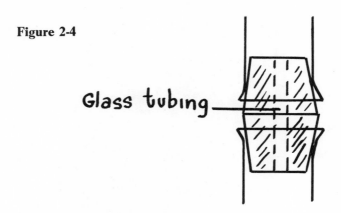

Glass tubing

5. Place the second rubber stopper into the top of the second test tube. Now invert the second test tube so that its one-hole stopper envelopes the 1-inch glass tubing of the first rubber stopper. The rubber stoppers are now back to back with the glass tubing in them (figure 2-5).

Have the students mark intervals of time on their sand clocks by looking at the seconds or minutes on the classroom clock.

PENDULUM AS A TIME DEVICE

The pendulum clock is based upon a device used for controlling the movement of clock work. This is a body suspended from a fixed point so as to move to and fro by the action of gravity and momentum.

Students can make simple pendulums with string and lead fishing sinkers. Many times problems may be proposed and solved by students in class. One of the questions asked by students was, "With the same weight, does the length of the pendulum have anything to do with the number of swings?"

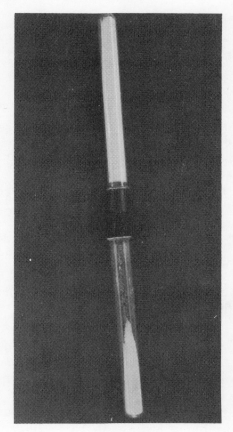

Figure 2-5

Making a pendulum

Materials:

String, small lead sinkers of the same weight, and a pencil.

Procedure:

1. Cut a piece of string 13 inches long.
2. Attach or tie one of the lead sinkers at the end of the string. Tie a small eye loop at the opposite end of the string.
3. Make a second pendulum by cutting a piece of string nine inches long and tying a lead sinker at the end of it.

Using the pendulum

The pendulums that the students have made are of two different lengths with the same amount of weight at the end.

Procedure:

1. Have the students work in small groups. Have them place a pencil through the upper loop of the longer pendulum. They then lift the weight parallel to the pencil, release it and count the number of swings it makes in 15 seconds.
2. They should conduct at least five trials, add up the swings per trial and take an average by dividing by five.
3. Repeat the above experiment using the shorter pendulum. You will find that the shorter pendulum will have more swings than the longer one.

Some students will ask, "Does the weight at the end of the pendulum have anything to do with the number swings when the length is the same?"

Procedure:

1. Have the students make two pendulums 13 inches long with one weight on the first and two equal weights on the second.
2. Each pendulum can then be tested again by counting the number of swings in 15 seconds. The students should conduct five trials as in the previous experiment, add them and divide by five to obtain an average for each pendulum.

They should find that the number of swings of the pendulum does not depend on the amount of weight at the end but only on its length (figure 2-6).

SCALE OF DISTANCES FOR THE PLANETS

The problem with working with the relative distances between the planets and the sun is the enormous distance between

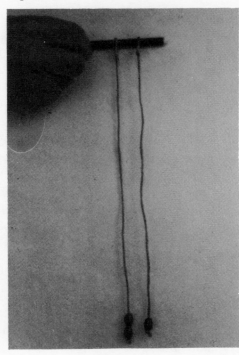

Figure 2-6

the one closest to the sun and the outermost. The following
scale of distances has been devised so that students may measure the relative distances of the planets within the classroom.

THE SOLAR SYSTEM — (1 INCH = 20,000,000 MILES)

Planet	Distance from Sun (in millions of miles)	Scale Distance from Sun (in inches)
1. Mercury	36	1 3/4"
2. Venus	67	4"
3. Earth	93	4 5/8"
4. Mars	141	7"
5. Jupiter	483	24 1/4"
6. Saturn	886	3' 3/4"
7. Uranus	1,782	7' 5"
8. Neptune	2,793	11' 6 1/2"
9. Pluto	3,680	15' 4"

Materials:

Yard sticks or rulers, adding-machine tape or strips of paper 2 inches wide, tape or paste.

Procedure:

Divide your class into small groups of two or three. Have them take strips of adding-machine tape or strips of paper and measure the scale of distances of the planets from the above scale, and in correct sequence from the sun from Mercury to Pluto. Write in the name of the planet at each step of the way. The strips of paper may be taped or pasted as additional strips are needed. The students should be able to measure all the planets on a strip of paper 15' 4" long.

SCALE MODEL OF THE PLANETS

This is a method of making a paper scale model of the planets. This activity has been done by individuals and in small groups of students. The scale used for the diameter of the planets was 1 inch equals 10,000 miles.

Materials:

Drawing compasses, unlined paper, pencils, string and scissors.

THE SOLAR SYSTEM (1 INCH = 10,000 MILES)

Planet	Diameter (Approx.)	Scale diameter (in inches)
1. Mercury	3,010	5/16"
2. Venus	7,575	3/4"
3. Earth	7,920	13/16"
4. Moon	2,160	1/4"
5. Mars	4,215	1/2"
6. Jupiter	86,800	9"
7. Saturn	72,500	7 1/4"
8. Uranus	30,800	3 1/8"
9. Neptune	32,900	3 7/16"
10. Pluto	4,000	1/2"
11. Sun	864,000	7' 2 1/2"

Procedure:

1. Give the students the above figures on the scale diameter of the planets.
2. Let them measure and make the circular planets with their compasses on sheets of unlined paper.
3. For Jupiter and Saturn they will have to make their own pencil compasses with string in order to draw so large a circle. The pencil compasses can be made by tying a string three-quarters of the way towards the point. Circles may be made by holding the string tied to the pencil down on the sheet of paper with the left hand and using the right hand with the pencil to make the circles.

The planets may be cut out and pasted or taped on the strip of paper of the scale of distances for the planets or may be hung on a separate sheet of paper.

WHAT YOU WOULD WEIGH ON THE PLANETS

In studying the planets and gravity students always ask how much they would weigh on a satellite such as the moon or other planets. By knowing the surface gravity of the planets and multiplying it by each student's weight, they will know how much they will weigh on a particular planet. The following activity was used for each student to figure out his weight on the different planets.

Procedure:

1. Have the students make the chart shown at the top of the next page.
2. The students would multiply their weight with the surface gravity of the planet in order to find out their weight on a particular planet or satellite. A student weighing a hundred pounds on earth would weigh only seventeen pounds on the moon. This gives the students a true perspective of how much they would weigh on the planets of the solar system.

SOLAR SYSTEM WEIGHT CHART		
Planet or Satellite	*Surface Gravity*	*Your weight*
Moon	.17	
Mercury	.36	
Venus	.86	
Earth	1.00	
Mars	.38	
Jupiter	2.64	
Saturn	1.17	
Uranus	.92	
Neptune	1.10	
Pluto	?	

PARTS OF AN AIRPLANE THAT HELP CONTROL ITS FLIGHT

When studying how an airplane flies and is controlled, it is not enough to label the parts that control the airplane, but to show the importance of aerodynamics in its control. It is therefore necessary to construct a paper model airplane that will have some type of moveable parts that will demonstrate these principles.

Designing the paper model airplane

In order for the students to perform the same uniform aerodynamic experiments and obtain similar results, it is necessary for the teacher to design a model paper airplane that can be duplicated on paper. A copy of this paper airplane is given to

each student. It is then cut out with scissors and the aerodynamic experiment is then carried out. After they have performed aerodynamic experiments they may wish to design their
own models.

Materials:

Duplicating paper (ditto) and a duplicating machine.

Procedure:

1. Make the wing 6 inches long and 1 inch wide.
2. Mark in the aileron on the right wing by measuring
   ¼ of an inch from the outer edge of the right hind
   wing. Make the ailerons 5/16 inches wide and 1 3/4
   inches long (figure 2-7).

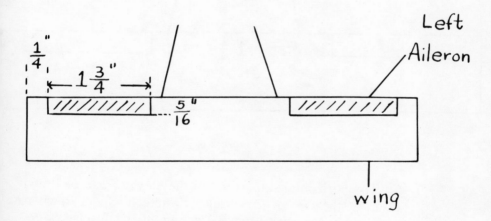

Figure 2-7

3. Mark in the aileron on the left wing by using the same
   measurements as in step two.
4. The body or fuselage of the airplane should be located
   in the center of the wing. Find the center of the wing
   with a ruler and measure 5/8 inches on each side of
   this point.

5. The body should extend back 2½ inches from the wing. The body should taper from 1¼ inches wide at the wing to ½ inch at the end (figure 2-8).

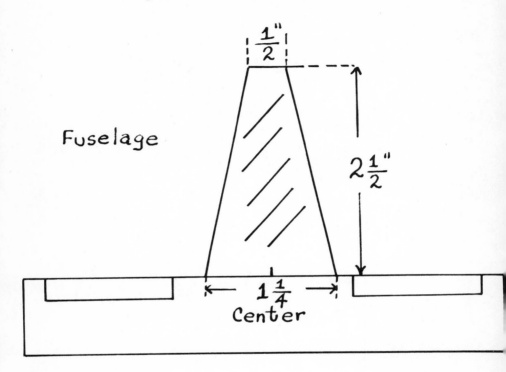

Figure 2-8

6. The horizontal stabilizer and the elevator can be drawn in at the end of the fuselage where it tapers down to ½ inch. Measure from the center of the fuselage's end and draw a line 1 1/8 inches to the right and left of this point or a total of 2¼ inches. Then extend it back 7/8 inches. At the end of this horizontal stabilizer measure 3/8 inches and draw a horizontal line. This will represent the elevator of the airplane (figure 2-9).

7. The vertical stabilizer and the rudder will have to be designed separately on the same paper. Draw a rectangle 3 1/8 inches long and 3/4 inches wide (figure 2-10).

Elevator

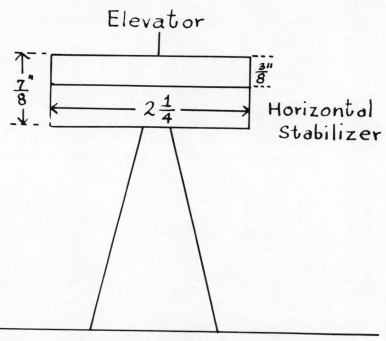

Horizontal Stabilizer

Figure 2-9

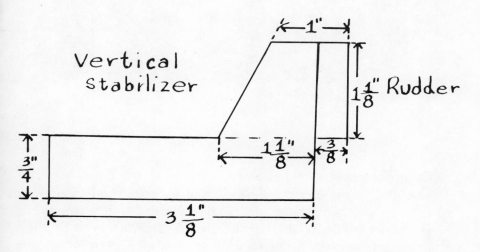

Vertical Stabilizer

Rudder

Figure 2-10

8. Measure and mark 1 1/8 inches from the right of the rectangle. This will be the width of the vertical stabilizer. Measure and mark 3/8 inches beyond the stabilizer. This will be the width of the rudder.

9. The vertical stabilizer and the rudder should be made 1 1/8 inches high. The stabilizer and rudder should be tapered at the top to about 1 inch (figure 2-10). Both drawings (the airplane's body and the stabilizer and rudder) should be on one sheet (figure 2-11).

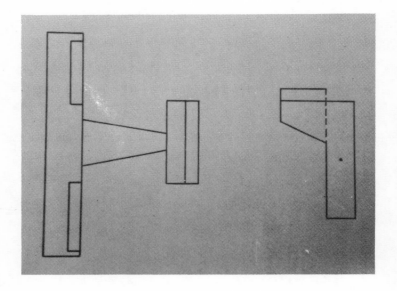

Figure 2-11

Using the paper model airplane

A ditto sheet with the airplane and its parts should be given to each student. The students should cut out the two parts. Have the students do the following experiments with their paper airplanes in order to discover how the parts of an airplane help to control it.

How ailerons control an airplane

In order to have the students find out how the ailerons control an airplane have them cut the edges of the ailerons. Have them bend the right aileron straight up and the left straight down (figure 2-12A).

Have them hold the airplane at the elevator or end with the index finger facing downward at about chin level. Let them drop the airplane about five times and have them record each time which way the airplane turns. The airplane should spiral or turn to the right. When the students have finished, they should then reverse the position of the ailerons. The right aileron should be bent downward and the left straight up (figure 2-12B). They should repeat the experiment, dropping the airplane five times and recording what happens each time. The airplane should now spiral or turn to the left.

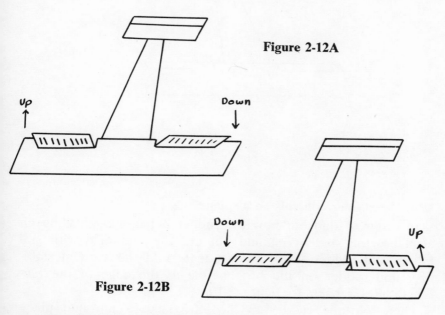

Figure 2-12A

Figure 2-12B

How the elevator controls an airplane

Have the students straighten out the ailerons and place the elevator at the end of the airplane in an upward position (figure 2-13).

Then have the students drop their airplanes again from chin level. Drop it five times and record what happens. When the airplane is dropped it should spiral upward and then fall to the ground.

The next part of the experiment is to reverse the elevator by bending it to a down position. Repeat the above procedure in dropping it and record what happens. The students should find that with the elevator in a bent down position, the airplane will spiral or turn downward.

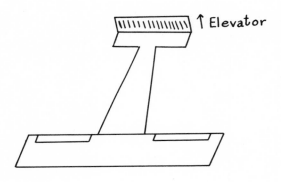

Figure 2-13

How the rudder controls an airplane

In order to find out how the rudder helps in controlling an airplane, the students should cut out the rectangle with stabilizer and rudder. Have the students fold the vertical stabilizer and rudder on the broken line at the edge of the rectangle. Refer back to figure 2-10.

Then fold the rudder to the left. Take a pin and place through the center of the rectangular piece of paper. With two fingers, hold the pin under the paper. The pin that the student is holding serves as a pivot for the movement of the stabilizer and rudder. Take a straw and blow against the rudder, bent left. The students will find that the rudder in a left position will move around to the right, but the body on the pivot pin will move to the left (figure 2-14A). After the students have recorded what happened when the rudder was in a left position, have them reverse the rudder to the right side (figure 2-14B).

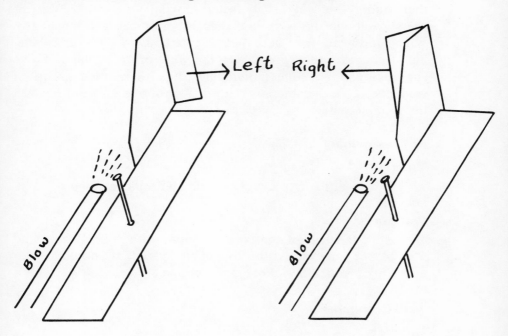

Figure 2-14A   **Figure 2-14B**

Have them blow with a straw against the rudder and record which way the body would turn. The rudder bent to the right would move to the left but would move the body of the airplane to the right.

When the students have completed these basic experiments indoors, they could design their own paper airplanes. When flight testing them outdoors, they may be judged by setting up categories such as maneuverability, distance or height.

How the wing lifts the airplane

The airfoil or wing of an airplane is curved on the top surface and straighter on the underside and is slightly tilted upward on the fuselage of the airplane. Lift is produced by unbalancing the air pressure on the wing by the forward motion of the airplane. Because the top of the wing is curved more, the air moving over the top of it travels a greater distance than on the underside. Therefore the air on top of the wing moves faster than under the wing. The faster moving air reduces the air pressure above the wing and thereby produces a lifting force.

The following four activities will demonstrate the unbalancing of air pressure to give lift to the wing of an airplane. These activities can be done individually or with a small group of students. Have the students record their observations of each of the four activities.

Activity number 1

Materials:

A strip of paper 1½ inches wide and 8 inches long.

Procedure:

Hold the end of the strip of paper with your fingers and blow gently across the top of the paper. It should rise or lift as you blow above the surface.

Activity number 2

Materials:

A 3 × 5 inch sheet of paper.

Procedure:

1. Bend 3 3/4 inches at the edges of both ends of the 3 × 5 inch sheet of paper.
2. Place the sheet on its edges at the end of a table and blow under it. The harder you blow the more the sheet will bow in the center. The moving air under the sheet has less pressure than the still air on top (figure 2-15).

Activity number 3

Materials:

Two mailing tubes 3½ inches long and a drinking straw.

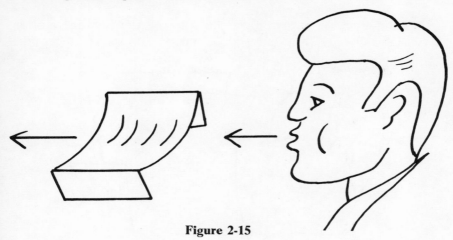

Figure 2-15

Procedure:

1. Place the two mailing tubes alongside each other about ¼ inch apart.
2. Take the drinking straw and blow down between the two mailing tubes. The mailing tubes should roll together. The blowing of air between the tubes reduced the air pressure, and the normal air pressure on the opposite sides forced them together.

Activity number 4

Materials:

Two sheets of binder paper.

Procedure:

Hold the two sheets up about two inches apart and blow between them. The moving air reduces the pressure and the two sheets should come together (figure 2-16).

How streamlining helps an airplane's flight

The next two activities will demonstrate how air flowing around an object will either increase its resistance (drag) or decrease it depending upon the shape of the object. These activities have been done by individuals or in small groups.

Figure 2-16

Activity number 1 — Demonstrating air resistance or drag

Materials:

Candle, 3 by 5 inch card, small dish.

Procedure:

1. Light a candle and melt a few drops in the center of a small dish. Place the end of the candle into the melted wax and hold it in position until the wax hardens.
2. The students should hold the 3 by 5 inch card upright a few inches from the flame.
3. Have them blow hard against the center of the card. The flame will move in the direction of the air current (figure 2-17). While you are blowing against the card, the flame bends towards it. This indicates that the air current going around the card is in a circular motion leading to the back of the card instead of flowing past it. It is this circular back flow that causes drag (the airplane losing speed).

Figure 2-17

Activity number 2 — Streamlining to increase the speed or decrease drag

Materials:

Candle, 5 by 8 inch card, small dish, masking tape or scotch tape.

Procedure:

1. Bend the ends of the 5 by 8 inch card towards each other and tape them together.
2. Light the candle and place it in the small dish.
3. Place the card with the taped ends a few inches from the candle.
4. Blow against the center of the curved surface of the cardboard. Have the students observe the direction the flame is pointing. It indicates the flow of the current of air. The flame should be back flowing with the air stream (figure 2-18). This shows that the streamlining or tapering of the body and wings helps increase speed (or decrease the drag).

Figure 2-18

HOW TO MAKE A ROCKET LAUNCHING PAD

Materials:

One piece of wood 1 5/8 by 3/8 inches, 20 inches long; one piece of wood 1 5/8 by 3½ inches, 19 inches long; two pieces of 2 by 2 inch wood one 18 inches long and the other 16; two pieces of ¼-inch plywood 18 inches square. An 18-inch piece of metal tubing ½ inch wide and a small 1½-inch hinge. A ½-inch long bolt.

Procedure:

1. Nail the 19-inch long piece of wood to the top left corner of the piece of wood 20 inches long. The two pieces nailed together take the appearance of an L-shaped structure (figure 2-19A).
2. Take one of the pieces of wood 18 inches long and nail it to the center of the bottom piece.
3. Nail the 16-inch piece in an upright position on top of the 18-inch piece (figure 2-19B).
4. Take the 18-inch piece of metal tubing and with a hammer smash ½ inch of one end together.

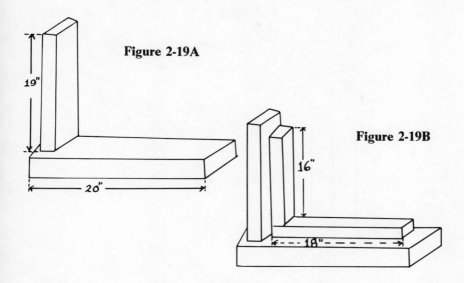

Figure 2-19A

Figure 2-19B

5. Make a hole through the center of the smashed end with a nail.
6. Take a ½-inch long bolt and pass it through the underside of the hole in the smashed tube.
7. Then place the bolt with the tube through the center hole of the 1½-inch hinge. Place a nut over the bolt on the opposite side (figure 2-20).

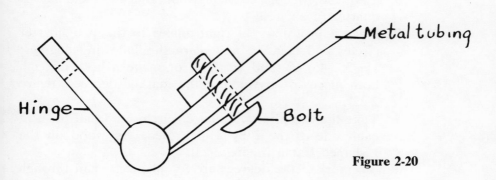

Figure 2-20

8. Nail or screw the opposite end of the hinge to the corner of the L-shaped 2 by 2 inch piece of wood (figure 2-21).

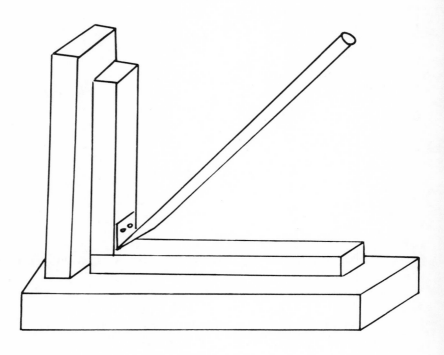

Figure 2-21

9. Take the two pieces of plywood and cut them into a quarter of a circle.
10. Place a protractor at right angles to the two quarter circles and measure and mark the following degrees: 15, 30, 45, 60, and 75. Drill or make a hole through both pieces of plywood at the marked degrees (figure 2-22).
11. Take the two plywood quarter circles and nail them to each side of the 2 by 2 inch pieces of wood on the L-shaped frame (figure 2-23).

The holes next to the degrees are for placing a nail through for holding the metal launching tube at the desired angle for launching a carbon dioxide rocket.

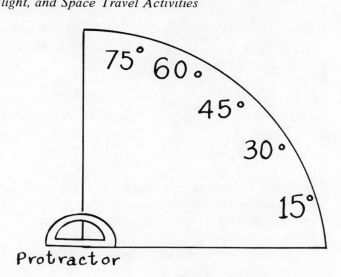

Figure 2-22

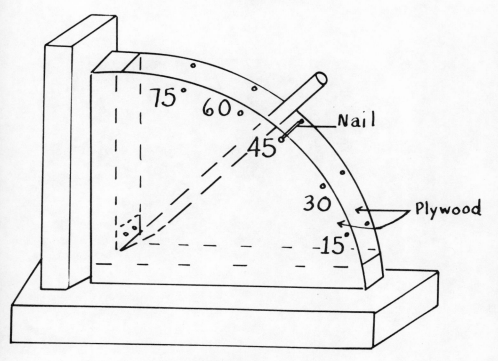

Figure 2-23

HOW TO MAKE A CARBON DIOXIDE ROCKET

This type of rocket is built to accompany the rocket launching pad described in the previous activity.

Materials:

   Carbon dioxide cartridges, ¼-inch wooden dowel and masking tape.

Procedure:

1. Measure the length of a carbon dioxide cartridge. Take the wooden dowel and measure approximately seven times the length or 21 inches of the carbon dioxide cylinder.
2. Tape the carbon dioxide cylinder lengthwise to the end of the wooden dowel with masking tape (figure 2-24).

Figure 2-24

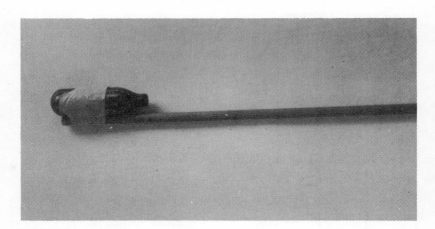

To launch the rocket, place it in the tube of the rocket launcher. The seal of the carbon dioxide cartridge may be punctured with a special Carbon Dioxide Puncture Gun when launching the rocket. The carbon dioxide gun can be purchased at a local hobby shop.

ROCKET DATA TEST SHEET

Before going out and firing the carbon dioxide rockets, you should develop with the help of the students a rocket data sheet. You could ask the students, "What possible data or information should be placed on the rocket test data sheet?" Place this list on the blackboard. The following information is pertinent: kind of fuel, weight of rocket, weight of fuel, launch date, launch angle, tangent angle, name or number of the rocket, flight speed, and rocket height.

The following Rocket Data Test Sheet might result:

ROCKET DATA TEST SHEET				
Kind of fuel	Fuel wt.	Rocket wt.	Launch date	Flight time

Name or number of rocket	Launch angle	Tang. angle	Range or Distance	Speed ft./sec.	Height ft.

The rocket weight was determined by simply weighing one of the full carbon dioxide cylinders on a balance scale. The weight of the rocket fuel or the carbon dioxide in the cylinder was determined by weighing an empty cylinder and subtracting that weight from a full one.

LAUNCHING THE CARBON DIOXIDE ROCKET

It is recommended to use a large field connected with the school grounds for the launching of the carbon dioxide rocket.

As a safety precaution make sure that the area in which the rocket is launched is cleared of students. Have the students stand behind the rocket launcher. Do not launch any rockets into the wind or on a gusty day. Launch them on a calm day with the wind or into a slight crosswind.

Procedure:

1. Place the rocket, the carbon dioxide cartridge taped to the 21-inch dowel, into the metal tube of the rocket launcher (figure 2-25).

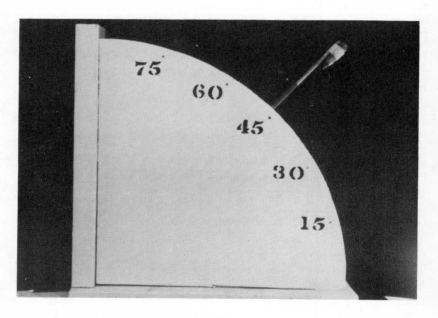

Figure 2-25

2. A carbon dioxide puncture gun may be used to break the seal in the back end of the carbon dioxide cylinder in order to launch the rocket. After the rocket is launched it may be retrieved and the used cartridge taken off the end of the stick. A new cartridge may be substituted and the rocket placed back into the rocket launcher for the next launching.

RECORDING DATA OF A ROCKET LAUNCHING

Before launching each rocket the class prepares a Rocket Data Test Sheet. The name or number given to the rocket and the launch angle and date should be recorded on the sheet before each launching. It was found that a 60-degree launch angle had the longest flight time and range.

Flight time

A stop watch or a watch with a second hand is needed to record the flight time of the rocket. The flight time is recorded in seconds from the time the rocket is launched to the time it touches the ground.

Range or distance

Some type of measuring device such as a tape measure is needed to measure the flight distance or range of the rocket. The flight distance or range is the distance from the launching pad to where the rocket hits the ground.

Calculating velocity (speed) of a rocket

With all the launch data recorded on the rocket data sheet, the speeds and the heights of the rockets launched can be calculated mathematically in the classroom. The velocity of a rocket can be calculated by dividing the flight time which is in seconds into the distance or range. The distance is measured in feet. The speed or velocity is then measured in feet per second.

Calculating height of a rocket

The height of a rocket can be calculated if the launch angle and the distance (range) are known. Take the distance measured in feet and divide this figure by two. Next obtain the value of the tangent angle for the launch angle of the rocket. Multiply this by one-half the distance (range) to obtain the height.

The target values for the following launch angles are:

Launch Angle	Tangent Value
30°	.57735
45°	1.0000
60°	1.7321

Additional tangent values can be found in most mathematics text books.

HOW A SATELLITE STAYS IN ORBIT

Whenever you study in class the topic of space, planets, or natural or artificial satellites an explanation of how a satellite stays in orbit is necessary. The satellite stays in orbit when the speed of the satellite is such that the centrifugal force (outward pull) balances with the force of gravity (downward pull). This principle cannot always be understood by the student unless he can participate in an activity that will demonstrate it. This activity will attempt to have the students build a device that will demonstrate how a satellite stays in orbit. This device when built should be demonstrated outdoors.

Materials:

Small 6 oz. juice can and string.

Procedure:

1. Make two small holes with a nail or point of scissors opposite each other near the top of the can.
2. Cut a piece of string about 7 inches long. Pass one end of the string through one of the holes and over the top of the lip of the can and tie it in place.
3. Take the other end of the string and tie it to the opposite hole. It should leave a string handle or an inverted "U" (figure 2-26).
4. Cut another 3 or 4 feet of string. Take one end of the string and tie it to the center of the string handle of the can.

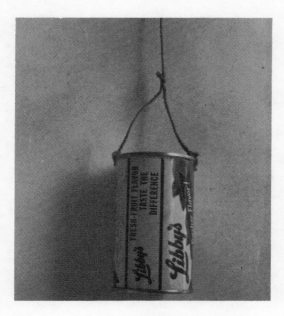

Figure 2-26

To demonstrate the principle of centrifugal force, take the device outdoors. Have the students fill the can three-quarters full of water. Have them hold the string near the can and slowly start spinning the can in a circular pattern around their heads while letting the string out through their fingers. The water should not fall if the centrifugal force (outward pull) is in balance with gravity (downward pull). They will notice that the faster it swings around their heads, the more the string stretches, and the greater the pull outward or centrifugal force.

— 3 —

CHEMISTRY
ACTIVITIES

Many of the abstracts in chemistry such as atoms, molecules and elements can become meaningful to students when visualized with the aid of models and charts. The first part of this chapter tries to accomplish this by building pie plate atoms, molecules, atomic slide rule, an electric quiz board of elements with symbols, and element blocks. The latter half of the chapter deals with making, discovering and testing invisible gases such as oxygen, hydrogen and carbon dioxide.

PIE PLATE MOBILE ATOMS

A variety of atoms can be built by students with pie plates. An atom of hydrogen will serve as an example in building the pie plate atom.

Materials:

Nine-inch pie plate, yarn or string, compass, ruler, 3 × 5 inch white card, scissors and paste.

Procedure:

1. Make two circles with a compass from the center of the pie plate. With a compass, draw the first circle 4 inches or 2 inches in diameter from the center of the plate. Draw the second circle 5½ inches or 2-3/4 inches in diameter (figure 3-1).

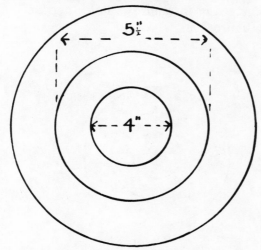

Figure 3-1

2. Cut out the 4-inch center circle with a pair of scissors. The center circle will be used as the center or the nucleus of the atom. Make small circles for each neutron and proton (+) to represent the particles in the center of the 4-inch circle.

3. Since hydrogen has only one proton (+) and no neutrons, you would make one circle with a (+) positive symbol in the 4-inch circle.

4. Cut out with scissors the 5½-inch circle. The large outside circle of the pie plate will represent the electron shells. Circular electrons with negative (−) charges can be drawn on the 3 × 5 white cards. They may be cut out and pasted on the large pie plate rim that represents the electron shells.

5. Since hydrogen has only one electron, you would draw one circle on a card and place a (−) negative symbol within it. Cut it out and paste it to the large rim of the pie plate.

6. Now you are going to make three small holes to place the string for your mobile atom. Make the first small hole in the top of the 4-inch circle that represents the nucleus of the atom.
7. Next make a small hole on both the outer and inner edge of the top of the rim that represents the electron shell.
8. Place the string through the holes and tie the atom together (figure 3-2).

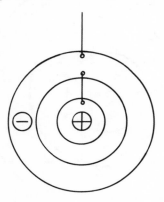

Figure 3-2

9. Cut the 3 × 5 card lengthwise and print the name of the atom, "Hydrogen". Paste the name card to the bottom of the atom. Hang up your mobile atom for display in the classroom (figure 3-3).

Figure 3-3

Each student may select different atoms to construct for mobiles. Other media used by students in the construction of atoms have been Styrofoam balls, pipe cleaners, yarn and wire, or a simple schematic model.

PIE PLATE MOBILE MOLECULES

One simple way to show how formulas or compounds are connected by molecules is to have students make pie plate molecules. In this activity, the compound methane will serve as an example.

Materials:

Two 9-inch pie plates, yarn or string, hand punch and a compass.

Procedure:

1. Take a pie plate and print the name and formula, such as methane CH_4 in the center. Punch a hole at the top and bottom of the pie plate with a hand puncher.
2. Take a second pie plate and draw with a compass one circle in the center. At the edge of the circle draw four more circles opposite one another. Each of the circles represents an atom that goes into the making of a molecule of methane. Place the letter "C" for carbon in the center circle with the letter "H" for hydrogen in each of the four outer circles. One atom of carbon and four of hydrogen make one molecule of methane. Punch a hole at the top of the second pie plate.
3. Take a 4-inch piece of yarn or string and pass it through the holes at the top of the second pie plate and at the bottom of the first. Make a large enough knot in the back of both holes so the knot won't slip through.
4. Cut another piece of yarn and pass it through the top hole of the first pie plate. Tie a knot in it. The molecular pie plates are now ready to be hung up as a mobile (figure 3-4).

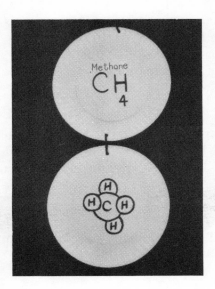

Figure 3-4

TINKERTOY MOLECULES

In this activity we will be building a molecule of water, H_2O.

Materials:

Tinkertoy set, 3 × 5 cards, masking tape, scissors, and a marking pencil.

Procedure:

1. Since a water molecule is made up of three atoms, you will need three circular disks or wooden knobs with ¼-inch holes in the side to make a molecule.
2. Place these circular knobs on a 3 × 5 card and trace them with a pencil.
3. Cut out the three circles with scissors and tape each on the flatside of a circular Tinkertoy wooden disk.
4. One molecule of water is made up of two atoms of hydrogen and one of oxygen. Take a marking pencil and print the letter "H" on two of the paper disks attached to Tinkertoy knobs and the letter "O" on the third.
5. In order to combine the three atoms, take two of the ¼-inch wooden dowels from the Tinkertoy set and

place them into the holes opposite each other in the atom with the symbol O. Connect the hydrogen (H) atoms to the ends of the two wooden dowels sticking out from the oxygen atom (figure 3-5).

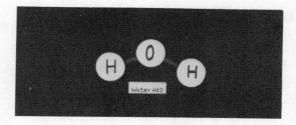

Figure 3-5

A small printed place tag with the name and formula "Water H_2O" can be placed with the molecule. These types of molecules could also be set up as mobile molecules with string or thread. Each student could select and construct different molecules. Figure 3-6 shows a few examples of Tinkertoy molecules that have been built by students.

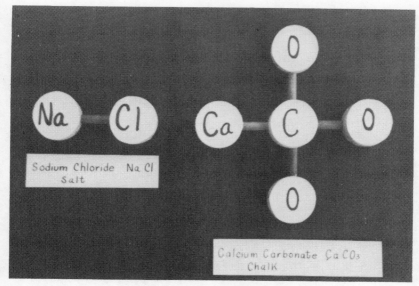

Figure 3-6

MAKING AND USING THE ATOMIC SLIDE RULE

This activity is a way of exposing students to learning how to interpret and use the information about particles of atoms with a simple atomic slide rule.

Making the atomic slide rule

Materials:

Two 5 × 8 cards, ruler and pencil.

Procedure:

1. Make a fold inward 1½ inches on each side of the card (figure 3-7A).
2. With the card folded measure ¼ inch on each side of the outer flap (figure 3-7B). Fold outward each flap along the ¼" line (figure 3-7C).

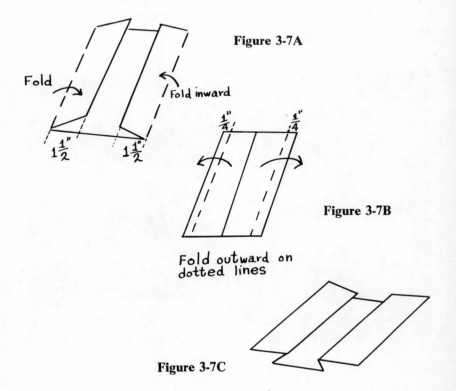

Figure 3-7A

Fold inward

Fold outward on dotted lines

Figure 3-7B

Figure 3-7C

3. At the top of the left flap print <u>No. of Electrons (−)</u>: at the top center of the card <u>Name   Symbol</u>; on the right-hand side <u>No. of Protons (+)   Neutrons</u> (figure 3-8).

No. of Electrons (−)	Name Symbol	No. of Protons Neutrons (+)

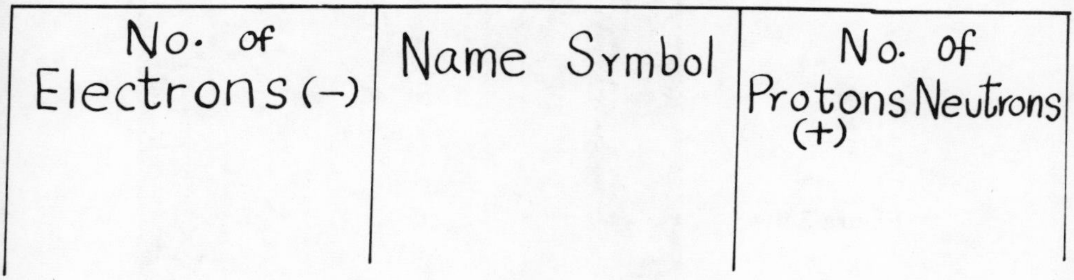

Figure 3-8

4. Select about eight elements and place the information about their atomic structure in each of the three categories on the card. Space the name of each element about 1 inch apart from the top to the bottom of the card. The eight lightest elements are used here. For example, hydrogen would be at the top of the card with the symbol "H". The number of electrons and protons would be 1, with zero neutrons (figure 3-9).

5. Take your second card and cut a strip 1 7/8″ by 8″. Print on the strip Atomic Slide Rule. Place this strip in the center under the grooves of your slide rule. The strip may be pulled down to any of the elements (figure 3-10).

How to interpret and use the atomic slide rule

Each student is to take a blank sheet of paper and make a diagram of the atomic structure of each of the elements on their atomic slide rule. I have chosen helium (He) and lithium (Li) as examples for making atomic diagrams from the Atomic Slide Rule (figure 3-11).

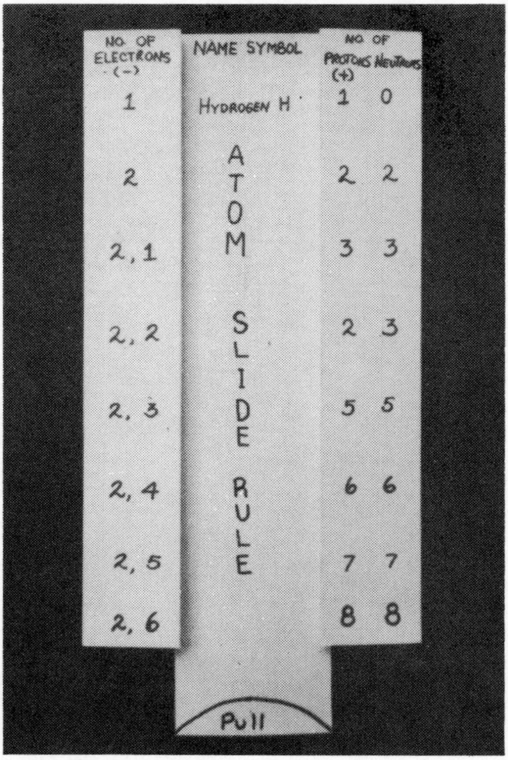

Figure 3-9

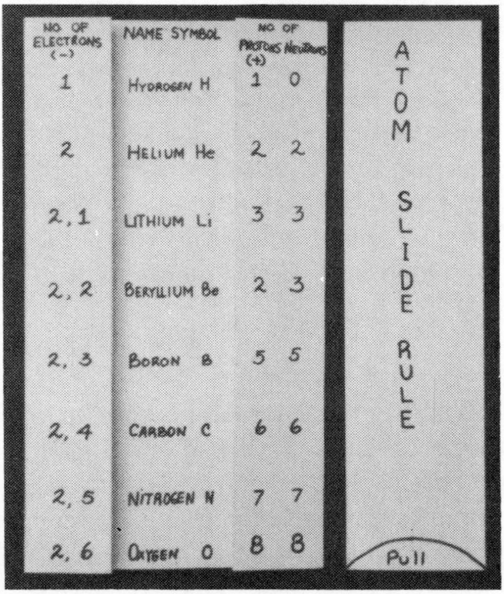

Figure 3-10

No. of Electrons (−)	Name Symbol	No. of Protons Neutrons (+)	
2	Helium He	2	2
2, 1	Lithium Li	3	3

Figure 3-11

Place the name helium and symbol "He" on a sheet of paper. Since helium has two protons, you would draw two circles alongside each other with a positive (+) symbol in them. Draw two more circles representing neutrons alongside each other next to the protons. You do not place any symbols in them because neutrons are neutral. The protons and neutrons make up the center or nucleus of the atom. To locate the number of outer particles or electrons (−), look to the left side of the slide rule under No. of Electrons (−). Since Helium is made up of two electrons, make two circles with a negative (−) symbol to represent the electrons on the outer edge (figure 3-12A).

Follow the same procedure in diagraming lithium. On the left side of the slide rule the electron arrangement in lithium is 2 and 1. The 2 stands for two electrons on the first orbital shell and the 1 stands for one electron on the second shell (figure 3-12B).

By taking the basic information about atomic particles from their atom slide rules, the students can make and interpret diagrams of different atoms.

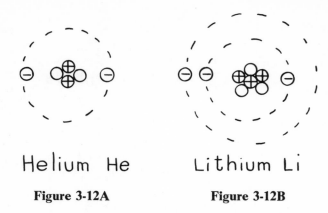

Helium He   Lithium Li

Figure 3-12A   **Figure 3-12B**

ELECTRIC QUIZ BOARD OF ELEMENTS
AND SYMBOLS

An interesting way to learn and review elements and symbols is with electric quiz boards. They can be built individually or by a small group of students. The device consists of a piece of cardboard carrying 3 $\times$ 5 cards with the names of elements on one side and an equal amount of cards with chemical symbols scrambled on the opposite side. A paper clip is clipped to a chemical symbol and a piece of wire extends from it to another paper clip attached to the correct element in back of the circuit board (figure 3-13).

By placing the electrical terminal of a light tester on an element and another one on the correct symbol, an electric light bulb goes on. The description of how to build the electric quiz board and the light tester is in chapter 1 under the subtitle, "How to Make an Electric Quiz board".

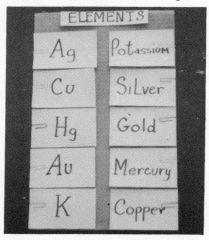

Figure 3-13

ELEMENT BLOCKS

This activity is appropriate when studying the elements or the periodic table of elements.

Building the element blocks

Materials:

One ½-gallon milk carton, 3 × 5 white cards or white paper, scissors, ruler and masking tape.

Procedure:

1. Take an empty ½-gallon carton and cut the top off.
2. With a ruler measure 3 3/4 inches from the bottom of the carton and mark it with a pencil. Draw straight horizontal lines at the 3 3/4-inch point (figure 3-14A).
3. Cut down at the four corners to the 3 3/4-inch line (figure 3-14B).
4. Measure another 3 3/4 inches on two opposite sides of the carton just above the first horizontal line (figure 3-14C). Cut both flaps at this point.
5. Measure and draw a line 1 inch above the first 3 3/4-inch horizontal line on the opposite flaps. Cut both flaps along the lines (figure 3-14D).

Figure 3-14A

Figure 3-14B

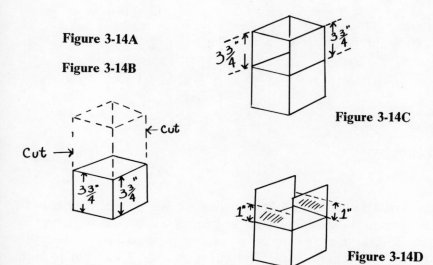

Figure 3-14C

Figure 3-14D

6. Fold the two smaller flaps inward and then the larger ones over them.
7. Tape the edges with masking tape. You will then have a cube or block.
8. Cover the outside of the cube with white paper or with plain white 3 × 5 inch index cards.

If ½-gallon cartons are not available, use quart or pint cartons.

How to use the element blocks

When the blocks are made, each student will select an element for individual study. The students place written information on the cube. The name and symbol of the element may be placed on the front. The atomic number (number of protons and the atomic weight (number of protons and neutrons) may be added.

On the top of the cube the atomic structure of the element may be drawn. General information about the element, such as its discovery, source, uses, and boiling and melting points may be placed on the back or side (figure 3-15).

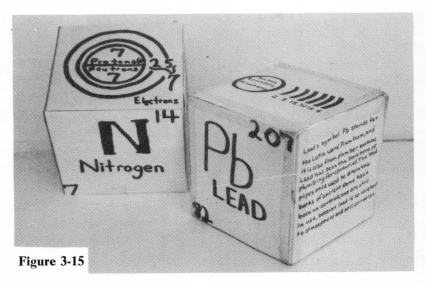

Figure 3-15

When the individual study cubes are completed, they may be combined by stacking them in general classes such as metals, non-metals, inert gases, or according to their atomic numbers or weights (figure 3-16). The cubes may be combined to form

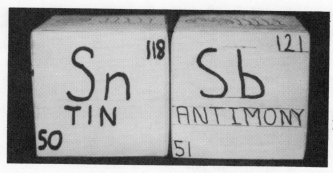

Figure 3-16

the periodic table of elements or they may be used by different students as a source of information for oral and written reports.

MAKING AND TESTING OXYGEN WITH HOUSEHOLD CHEMICALS

The activity of making and testing the vital gas oxygen has been carried on in the classroom for many years with great success. Part of the success came from the use of household chemicals. These chemicals were safe and inexpensive. It was found that empty baby food jars made excellent containers for mixing the chemicals and for testing the gas with a flaming wooden splint. Student test tubes were too small and confining for mixing and testing the gas with a flame, but were an aid in measuring and pouring the chemicals into the baby food jars.

Materials:

Hydrogen Peroxide $2H_2O_2$ (3% solution), Sodium hypochlorite $NaClO + H_2O$ (laundry bleach, Clorox brand), wooden splints, wooden matches, test tube, and a baby food jar. Safety glasses if available.

Procedure:

1. Fill a test tube 1/3 full of bleach and pour it into the jar. Rinse the test tube with water.
2. Fill the test tube 1/3 full of hydrogen peroxide. Before pouring the hydrogen peroxide into the jar, light a wooden splint or a wooden match.

3. Pour the hydrogen peroxide into the jar with the bleach. Hydrogen peroxide is used because pure oxygen can be liberated when you throw in a catalyst such as bleach. Hydrogen peroxide ($2H_2O_2$) is made of two molecules each of which contains 2 hydrogen atoms (H_2) and 2 oxygen atoms (O_2). When a catalyst is poured into hydrogen peroxide, oxygen (O_2) is liberated and water ($2H_2O$) is formed.

$$2H_2O_2 \longrightarrow O_2 + 2H_2O$$
hydrogen peroxide   oxygen   water

4. While the solution is fizzing, place the burning wooden splint or match in the jar just above the fizzing, where the gas is being released. You will notice that the flame will grow large and will glow brighter while the gas is being released (figure 3-17).

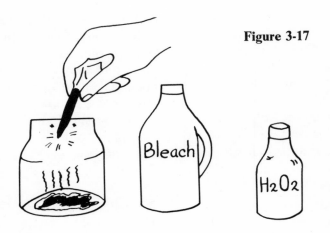

Figure 3-17

5. Repeat the experiment above. This time light the wooden splint or match, then blow the flame out but permit the red embers on the wooden splint to remain glowing.

6. Place the splint with the embers into the jar that is releasing the oxygen gas. The wooden splint should ignite. If your splint didn't ignite it's because the burning embers on the wooden splint went out. Make sure

that there is a good burning glow of embers before placing it into the gas.

PERCENTAGE OF OXYGEN IN THE AIR

This experiment has been done as a teacher demonstration and in small groups by dividing the class into five or six groups. While studying oxygen gas in class, some students will always ask, "How do we know what percentage of the air is oxygen?" This is the time for the teacher to say, "Let's see if we can figure it out."

Materials:

Pie pan or large finger bowl, 1 candle, matches, gas-collecting bottle or a quart jar, ruler and water.

Procedure:

1. Light a candle. Drip some of the melted wax in the center of a pie pan or finger bowl and fasten the candle to it.
2. Fill the pie pan 3/4 full of water.
3. Re-light the candle and let it burn for a few seconds until there is a large flame.
4. Carefully lower the bottle over the candle into the water and onto the finger bowl.

As the candle burns and goes out, the water from the pan will go up into the jar and take the space of the oxygen that was used by the candle (figure 3-18). (Oxygen occupies 21% of the air.)

Figure 3-18

Now the problem is to have the students try to figure out the percentage of oxygen in the bottle. Have the students measure the height and the level of water in the bottle. Then they can give an approximate percentage of the amount of oxygen in the bottle. Most groups of students come close in their approximations. On the basis of their observations, they usually estimate 20 to 25%.

HOW OXYGEN COMBINES WITH MATERIALS — OXIDATION

Rapid oxidation such as lighting a match can easily be seen. Rusting of iron or steel is a process of slow oxidation, whereby oxygen combines with steel at a slower pace. This experiment will demonstrate slow oxidation.

Materials:

Two test tubes, steel wool, beaker and water.

Procedure:

1. Take about one square inch of steel wool and roll it loosely into a ball between the palms of your hands.
2. Soak the steel wool in water for a few seconds and squeeze out the excess water. Do not compact the steel wool too tightly.
3. Pack the steel wool at the bottom of a test tube with a pencil. The steel wool should remain in the bottom when the test tube is placed upside down.
4. Place about 1 or 2 inches of water in a beaker.
5. Lean the test tubes (one empty) upside down against the inside of the beaker alongside each other. The level of the water in both test tubes should be equal. The empty test tube serves as a comparison or control.

Have the students make a sketch of the apparatus and record the date and the level of the water in each test tube. Let the apparatus stand undisturbed for 48 to 72 hours or until rust forms. The water level in the test tube with the steel wool will rise before the rust is visible. Have the students record the water line by shading in the area on their sketches (figure 3-19). Ask the students to try to explain why a second test tube was used and why the level of water was higher in the test tube with the steel wool.

Figure 3-19

Have them look for rust spots on the steel wool in the test tube. If they can't see any, have them remove the steel wool from the test tube and wipe it with a paper towel. Check it for signs of rust. The higher level of water in the test tube with steel wool shows that oxygen from the air joined with the steel wool and the water took the place of the oxygen. The rust indicates that oxygen has joined the iron in the steel wool, forming a compound, iron oxide:

$$4Fe + 3O_2 \longrightarrow 2Fe_2O_3$$
iron oxygen iron oxide (rust)

This demonstrates a form of slow oxidation.

HOW TO MAKE AND TEST FOR HYDROGEN GAS

Hydrogen is the lightest of all the gases. Hydrogen burns so quickly that it explodes. This is why hydrogen is safe to make only in small quantities.

Materials:

Vinegar ($HC_2H_3O_2$), 1 test tube, magnesium metal ribbon (Mg), matches, 1 wooden splint, a spoon and safety glasses if available.

Procedure:

1. Place two spoonfuls of vinegar into a test tube.
2. Drop ¼-inch piece of magnesium ribbon into the vinegar. You will notice a light fizzing around the magnesium strip. This is the release of hydrogen gas.
3. Quickly place your thumb over the test tube and hold for about two and a half minutes (figure 3-20A).
4. Light a wooden splint and tilt the test tube slightly away from yourself and partner. Remove your thumb from the test tube and quickly place the lighted wooden splint at the open end. As you do so, there will be a flashing blue flame with a "popping" sound as the hydrogen burns (figure 3-20B). If nothing happens your gas has probably already escaped. Therefore, you will have to repeat the experiment.

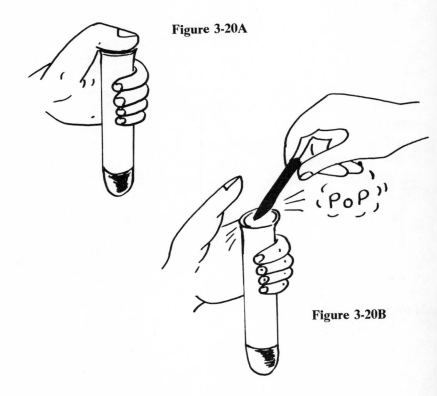

Figure 3-20A

Figure 3-20B

HOW TO MAKE CARBON DIOXIDE GAS

Carbon dioxide gas is one of the gases of the air. The gas can be made easily in quantity in the classroom with household chemicals.

Materials:

Acetic acid ($HC_2H_3O_2$) (vinegar), sodium bicarbonate ($NaHCO_3$) (baking soda), test tube, small jar and a teaspoon.

Procedure:

1. Place a teaspoon of sodium bicarbonate into a small jar.
2. Add about half a test tube of vinegar over the baking soda in the jar.

You will notice that the solution will fizz and make bubbles. This is carbon dioxide gas. You may ask your students to make a list of observational questions about the gas on a sheet of paper or on the blackboard. Some of the sample questions that might be asked are: "Does the gas have any color?" or "How do you know the gas is being released?"

HOW CARBON DIOXIDE REACTS TO BURNING MATERIALS

Materials:

One candle, matches, 1 jar with a screw-on lid, acetic acid ($HC_2H_3O_2$) (vinegar), sodium bicarbonate ($NaHCO_3$) (baking soda), a wooden splint, and safety glasses if available.

Procedure:

1. Light a candle and mix vinegar and sodium bicarbonate as described in the previous experiment. Screw the lid on the jar while the vinegar and baking soda are fizzing. Unscrew the lid and pour the gas, not the solution, over the lighted candle. The gas from the jar should extinguish the flame of the candle. If it doesn't, repeat the experiment.
2. Repeat the experiment above without the candle.

Instead use a lighted match or wooden splint. When the carbon dioxide is being made in the jar and is fizzing, unscrew the lid and place the lighted match or wooden splint into the jar. It should go out.

The students will discover that this gas is not only invisible but also puts out fire.

HOW TO MAKE LIMEWATER

When limewater is added to carbon dioxide gas, the liquid becomes cloudy or milky. Limewater then is used as an indicator for carbon dioxide gas.

Materials:

Calcium hydroxide [Ca (OH)$_2$] (slaked lime), 1 pint or quart jar, a teaspoon, and safety glasses if available.

Procedure:

1. Place a teaspoon of calcium hydroxide (slaked lime) into a pint or a quart jar of water. Cap the jar, shake the solution and let it sit overnight. The lime will settle to the bottom of the jar.
2. Pour out the solution when needed. If the solution is not clear enough, filter it through a paper filter or towel.

HOW TO DETECT CARBON DIOXIDE
WITH LIMEWATER

Limewater is an indicator for carbon dioxide gas.

Materials:

Limewater, acetic acid (HC$_2$H$_3$O$_2$) (vinegar), sodium bicarbonate (NaHCO$_3$) (baking soda), two test tubes and a teaspoon.

Procedure:

1. Place ¼ of a teaspoon of sodium bicarbonate into a test tube.

2. Pour slowly about ¼ of a test tube of vinegar into the test tube containing sodium bicarbonate. As the solution fizzes, place another test tube upside down over the one releasing carbon dioxide gas. Tilt both test tubes slightly so the gas can flow from the one where it is generated into the empty test tube (figure 3-21). The tilting is necessary to make the gas flow because carbon dioxide is heavier than air.

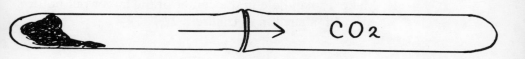

Figure 3-21

3. When the solution stops fizzing, place your thumb over the test tube in which you have collected the carbon dioxide gas.
4. Release your thumb and pour in about 1 teaspoon of limewater. Place your thumb back on the mouth of the test tube and shake it vigorously up and down for a few seconds. Look at the limewater solution. The limewater should turn cloudy or milky to indicate the presence of carbon dioxide gas.

HOW TO MAKE A MINI-PILL BOTTLE FIRE EXTINGUISHER

The problem of making a small functional fire extinguisher has been to find the quantity of materials for an entire class or many classes that would illustrate the principle of a soda-acid carbon dioxide fire extinguisher. Pill bottles and small vials were found easy for students to collect or they could be purchased inexpensively at local drugstores. In a soda-acid carbon dioxide fire extinguisher there is a small cylinder within a larger one. The smaller one is filled with an acid while the larger one is filled with a sodium bicarbonate solution.

Materials:

One pill bottle, 1 small vial that fits into pill bottle, piece of flexible wire, sodium bicarbonate, vinegar, small nail, and a teaspoon.

Procedure:

1. Make a small hole in the center of the larger pill bottle's cap with a small nail (figure 3-22).

Figure 3-22

2. Wrap a small piece of wire around the top lip of the small vial that will fit into the larger pill bottle and twist the wire leaving a 1½-inch leader of wire (figure 3-23).

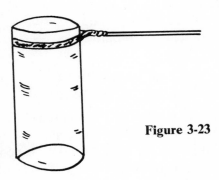

Figure 3-23

3. Fill the small vial with vinegar.
4. Place one teaspoon of sodium bicarbonate into one half of a glass of water and stir it with a spoon until dissolved. Fill the larger pill bottle about 3/4 full of the soda solution (figure 3-24).

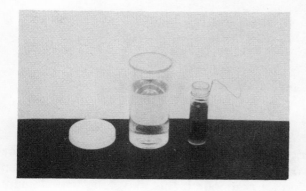

Figure 3-24

5. Lower the small vial with the vinegar into the pill bottle with the soda solution. Make certain that the soda solution does not touch the vinegar. If there is too much soda solution, lift out the smaller vial and pour out some of the soda solution. Place the smaller vial back into the larger one, making sure that the soda solution level is lower than top of the vinegar vial. Bend the 1½-inch leader wire over the lip of the larger pill bottle.
6. Place the cap with the small hole over the top of the vial with the soda solution.

In order for the mini-extinguisher to operate, tilt it down at an angle so that the vinegar and soda solutions will mix. With your fingers press the cap tightly against the pill bottle because the pressure of the carbon dioxide may force it off. The carbon dioxide gas will force the liquid out of the hole in the cap (figure 3-25).

Figure 3-25

HOW TO MAKE A COMPOUND

This activity demonstrates two principles. One is that the mixing of two elements does not mean that they will chemically combine to form a compound. But, the same two elements can be chemically combined into a compound by heating them.

Materials:

Sulfur (S), iron filings (Fe), alcohol lamp, spoon, beaker or petri dish, bar magnet, and safety glasses if available.

Procedure:

1. Measure out about 1/3 teaspoon of sulfur and 1/3 teaspoon of iron filings and place them in a petri dish.
2. Mix the two elements. Look and see if you still can identify the sulfur and filings.
3. Dip a magnet into the mixture. You will notice that the iron filings will cling to the magnet and can be separated from the mixture (figure 3-26).
4. Re-mix the sulfur and iron filings again in the petri dish.
5. Light the alcohol lamp and heat a spoonful of the mixture until it glows and melts.
6. Allow the contents in the spoon to cool. Examine the contents to see if you still can identify the sulfur and

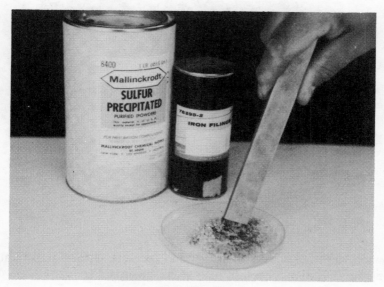

Figure 3-26

iron filings. Touch a magnet to the substance. You will notice that it will not cling to the magnet (figure 3-27). This shows that the substance is now a different material. A molecule of iron (Fe) has combined with a molecule of sulfur (S) to form a compound, iron sulfide (FeS).

$$Fe + S \longrightarrow FeS$$
iron sulfur   iron sulfide

Make sure that the iron and sulfur are heated sufficiently, otherwise there will be no chemical combination.

Students may take their samples of iron sulfide and make a visual display showing how the elements combined into a compound.

HOW TO BREAK A COMPOUND INTO ELEMENTS

One of the compounds that can be broken down into elements is mercuric oxide ($2HgO$). When heated the compound is broken

Figure 3-27

down into the elements mercury and oxygen. This experiment is done by dividing the class into small groups.

Materials:

Mercuric oxide (2HgO), test tube, alcohol lamp, test tube holder, wooden splints and a teaspoon.

Procedure:

1. Place 1/8 to a ¼ of a teaspoon of mercuric oxide into a test tube.
2. Light the alcohol lamp and heat the test tube with a test tube holder for a few minutes.
3. While one student is heating the mercuric oxide, have another student light a wooden splint. Place the glowing splint into the opening of the test tube. Have the students record their observations on a sheet of paper. The splint should glow or burn brighter. The glowing or brighter splint indicates the release of oxygen gas from the mercuric oxide.
4. Have the students examine the silvery ring in the test tube. They will notice small silver droplets clinging to the inside of the test tube. These are the mercury that came out of the compound. When mercuric oxide

(2HgO) was heated, mercury (Hg) and oxygen (O_2) was released. Therefore a compound was broken into two elements:

2HgO $\longrightarrow$ 2Hg and O_2
mercuric oxide   mercury   oxygen

CHEMICAL INDICATORS

Making red cabbage indicator

A boiled red cabbage solution is a good indicator for a base and an acid. Red cabbage leaves are given to students. They take the leaves home and boil them in water and bring the cabbage solution next day for experimentation in class.

Materials:

Red cabbage and water.

Procedure:

1. Cut a leaf or red cabbage into small pieces and add it to about ½ quart of boiling water. Boil for approximately 15 minutes. The water will turn purple or bluish.
2. When an acid such as clear vinegar is added to the cabbage solution, it will change to a red or dark pink. This indicates the presence of an acid.
3. When a base such as household ammonia is mixed in the cabbage solution, the color will change to green.
4. When an acid such as clear vinegar is slowly added to the green solution, it will become clear. This indicates the solution is now neutral. If you continue to pour vinegar, the solution will turn back to pink indicating an acid solution.

Students may make cabbage indicator strips by cutting filter paper into strips, soaking them in the solution and letting them dry overnight. The strips can be used to test chemicals and foods to see if they are acid or base. The strips also will turn green when placed in a base solution and red or pink in the presence of an acid.

Making phenolphthalein indicator

Phenolphthalein is an indicator for bases. Small amounts in powder form may be purchased from a drugstore.

Materials:

Phenolphthalein powder and denatured alcohol (rubbing alcohol).

Procedure:

Mix a pinch of phenolphthalein powder into 50 ml. (2 oz.) of denatured alcohol. Shake or stir the alcohol until the powder is dissolved.

Add a few drops of the solution to a base such as ammonia, limewater or baking soda solution until it turns red. When you add an acid such as vinegar to the solution and stir it, it will become colorless. If you now add a base, it will change back to red. The color may be changed back and forth by adding a base and then an acid.

Limewater indicator

To see how to make limewater for testing carbon dioxide gas, refer in this chapter to "How to Make Limewater."

— 4 —

ACTIVITIES THAT
HELP TO EXPLAIN
THE HUMAN BODY

Students are interested in the study of how the human body is built and how it works because of the great physical changes taking place in their own bodies. The organization of this chapter is designed to take advantage of the students' interest in the structure and function of the body in relationship to health and growth development. Many activities such as "Determining the length of the alimentary canal" are planned to catch pupils' interest and help them identify with it.

Some experiments such as "How saliva helps in the digestion of food," and "Pulse beat" demonstrate what kind of changes take place in the body. Some instruments such as "The spirometer" or "The pill bottle model of the lung" will measure or demonstrate some of the changes in the body.

THE MICROSCOPE

The microscope is a valuable tool for discovery in the classroom. It is generally a good idea to have students familiarize themselves with the parts of the microscope before using it. One method of helping them is to trace the parts and their names on a ditto sheet and duplicate (figure 4-1).

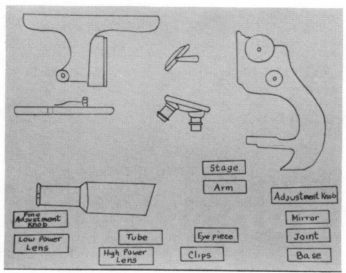

Figure 4-1

By referring to a chart or science book that has a picture of the microscope, students can cut out the six parts from the ditto sheet. These parts are then put together on a blank sheet of paper as in a jig-saw to form the correct shape of the microscope. When the student thinks he has the microscope put together correctly, he calls the teacher to check. He then pastes the microscope parts together on the blank sheet, cuts out labels and places them next to the parts of the microscope. The students can ask the teacher to check the labels before pasting them in place. The result will be a picture of a typical microscope (figure 4-2).

The following uses of the parts may also be written in by the students:

Eyepiece — Lens for magnification.

Tube — Hollow cylinder, forms the body of the microscope.

Adjustment Knob — Raises and lowers the body tube for focusing.

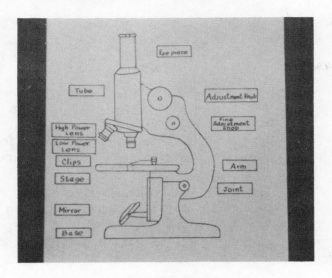

Figure 4-2

Fine Adjustment Knob — Used for sharp focusing.
High- and Low-power Lenses — Series of lenses, from lower to higher magnification.
Arm — A place to grip the microscope while carrying it.
Clips — Hold the microscope slide in place.
Stage — A platform upon which is placed the material to be observed.
Mirror — To direct light through the lenses.
Joint — To bend the microscope.
Base — U-shaped to support the microscope.

Practice with the microscope

Steps for students to follow in using the microscope are:

1. Bend the microscope to the position that will be comfortable to look through the eyepiece.
2. Wipe lenses with clean cloth or lens paper.
3. Click or place the low-power lens in line with the tube.

4. Look through the eyepiece and turn the mirror towards a light source such as a window to let in as much light as possible. Never point the mirror towards direct sunlight.
5. In focusing, always start with the low-power lens and lower it as close as possible to the thing you are going to study without touching it.
6. Then move the low-power lens slowly upward from the slide, using the focusing knobs, until it comes into focus.
7. When you are finished with the microscope, remove the slide, wipe off the stage, replace the clips in proper position and swing the low-power lens into position.

Now the students are ready to make and look at a practice microscope slide.

Materials:

Microscope slides, cover slides, eyedroppers, wool or silk cloth.

Procedure:

1. Place a strand of cloth fiber in the center of a microscope slide.
2. With the eyedropper place a drop of water over the strand of cloth.
3. Cover the drop and the strand with a cover slide.
4. Place the slide under the clips in the microscope.
5. Have the students observe the fiber and make two practice drawings of it under low- and high-power lenses.

EXAMINING YOUR CELLS

This is a simple way to have students examine their skin cells.

Materials:

Microscope, slides and cover slide, flat tooth picks, and tincture of iodine.

Procedure:

1. In order to obtain skin cells, have the students scrape the inner lining of their mouth with a toothpick.
2. Carefully place the moist tooth pick with the cells in the center of a microscope slide.
3. Place a drop of iodine over the cells and cover them with a cover slide.
4. Examine the slide under the microscope.

The students should be able to see typical skin cells with the cell membrane and a dark spot in the center which is the nucleus. Make a drawing and label the parts of the cell.

DETERMINING THE LENGTH OF THE ALIMENTARY CANAL

When studying the alimentary canal or digestive system a person learns the names of the organs but does not visualize its total length. This activity is to have the students measure the length of an average adult's alimentary canal and cut out and paste drawings of it. This activity will be more likely to succeed if the students are familiar with the organs that make up the alimentary system. It should be conducted in groups of two or three students.

Materials:

Unlined paper, rulers, paste and scissors.

Procedure:

1. The first part of the alimentary system to be measured is the esophagus or food tube. The esophagus is about 10 inches long and 1½ inches wide. Have the students measure, cut out, and label a paper representation of the esophagus.
2. The next organ in line is the stomach. Draw a sack-like structure 12 inches long and 4½ inches wide. Cut out and label the stomach and paste it to the esophagus.
3. The next structure leading down the alimentary canal is

the small intestines. Measure and cut out strips of paper
1½ inches wide. Paste the strips together until you have
the total length of 23 feet that makes up the small intestines. Paste the total length of the small intestines to
the end of the stomach.

4. The last part of the digestive system is the large intestines. Measure strips of paper 2½ inches wide and
paste them together for a total length of 5 feet to make
up the large intestines. Attach the large intestines to the
end of the small intestines. Additional accessory organs
such as the pancreas, liver, and gallbladder may be
drawn in or placed in the digestive system. The approximate measurements of these accessory organs are
as follows:

   Pancreas — (Tapers to a tail) About 5 inches long
   and 2 inches wide.

   Liver — 8 to 9 inches long. Width is 6 to 7 inches
   at the thickest part and tapers down to 4 to 5 inches.

HOW SALIVA HELPS IN THE DIGESTION
OF FOODS

In order to test the effects saliva has on the digestion of food,
you need a food known to have sugar and another that has
starch as a control or comparison. I would suggest a piece of
apple for the food with sugar and cornstarch for the food with
starch.

Materials:

Apple, cornstarch, Benedict's solution, saliva, three test
tubes, stirring rod, and a spoon.

Procedure:

1. Place a pinch of cornstarch and about a teaspoon of
saliva in a test tube and mix them with a stirring rod.
2. In the second test tube place a pinch of cornstarch and
about one teaspoon of water and mix them with a
stirring rod.

3. Place a small piece of apple and about one teaspoon of water in the third test tube.
4. Add one-quarter test tube of Benedict's solution to each of the three test tubes.
5. Heat each one of the test tubes and record the color changes. The starch and water solution remains blue. The apple and water change to a greenish yellow indicating some sugar. The saliva and starch change to a greenish color. The starch and saliva solution is closer in color to the apple solution than to the starch and water one, which demonstrates that the enzyme ptyalin helps to change starch to sugar.

TESTING FOODS

Testing food for starch

A simple way of determining if a food contains starch is to use the iodine test for starch.

Materials:

Tincture of iodine, bread, potato, cornstarch and water solution, test tubes. Control: apple and carrot.

Procedure:

1. Have the students make a list of possible foods such as above to be tested for starch.
2. Then add a few drops of tincture of iodine to each of the foods to be tested. Have them record the color changes of each food. If the food tested contains starch, it will change to a blue-black. Use foods such as apples and carrots as a control. They do not contain starch, therefore they will not change colors.

The students should bring to class as many different foods as possible from home to be tested.

Testing food for sugar

Materials:

Fehling's or Benedict's solution, apple, raisin, test tubes and test tube holder. Controls: cornstarch and potato.

Procedure:

1. Place a small piece of apple in a test tube.
2. Add a small amount of Fehling's or Benedict's solution and heat. If the solution turns an orange-red, this indicates a high amount of sugar. A greenish-yellow indicates a lower amount of sugar. If the solution remains blue, this indicates that no sugar is present. Table sugar will not react with Benedict's solution.
3. Have the students test some materials that do not have sugar in them, using cornstarch and potato as a control.

Testing foods for fats

Materials:

Unglazed paper such as tracing paper, bacon, butter, lard, oil. Control: bread, soda cracker and carrot.

Procedure:

1. Rub a piece of bacon or butter on a piece of tracing paper.
2. Hold up the piece of paper to the light. If you see a translucent spot or stain, this indicates the presence of fat.
3. As a control rub a piece of bread and a carrot on the same sheet of paper. Hold up the paper to the light and compare it to the bacon and butter stains. The students will notice that it does not leave a permanent stain on the paper.

Have the students make a list of foods to be tested and record the results.

Testing foods for proteins

Because of the possible dangers of the chemicals used for this experiment and the age level of the students, this activity should be done by the teacher as a demonstration for the class.

Materials:

Nitric acid, ammonium hydroxide, white of a hard boiled egg, milk and test tubes. Control: potato and granulated sugar.

Procedure:

1. Place a few pieces of egg white in a test tube.
2. Add a little nitric acid and heat for a few minutes. Do not boil it. If the chemicals spill on your hand or clothing, wash quickly with water.
3. Pour out the nitric acid and rinse the egg whites with fresh water. Add a small amount of ammonium hydroxide. Notice the color will be orange-yellow or yellow. This orange-yellow or yellow indicates the presence of protein. The paler the yellow, the less protein a food has.

A possible class project might be to test one of the school's lunch menus for starches and sugars and record the results in the school newspaper.

Testing for vitamin C

Citrus fruits such as oranges, grapefruits and lemons are rich in vitamin C (ascorbic acid). The vitamin is stored in the body in only small amounts. It is important for people to eat some source of vitamin C daily in order to meet the body's daily need. Vitamin C is essential for growth, strength of blood vessels, teeth development, gum health and prevention of scurvy.

Materials:

Orange juice, starch, water, iodine and a beaker.

Procedure:

1. Place a teaspoon of starch in a beaker of water and stir the solution.
2. Fill a test tube about one-third full of the starch solution and add one drop of iodine. The starch solution will turn blue-black.
3. Add orange juice drop by drop until the solution clears. This color change indicates the presence of vitamin C.

One project that might be done would be to test different types of oranges such as fresh, frozen, and instant dehydrated brands for varying amounts of vitamin C. The fewest drops needed to make the starch solution clear indicates the brand with the most vitamin C.

PULSE BEAT

Each time the heart beats, blood is pumped into the arteries in a wave-like motion. This is known as a pulse. A student can count the number of heart beats by feeling the pulse. The number of waves he feels when he places his fingertips on the wrist artery indicates the number of heart beats. Have the students find the pulse in their left wrist by using the first two fingers of the right hand. Do not use the thumb.

The following activity might be set up by the teacher on the blackboard or as a student worksheet.

Problem:

What is my pulse beat?

Procedure:

Take your pulse beat four separate times for one minute each. Find your average pulse beat by adding up all the figures and dividing by four.

Observation:

    My count 1st time ————————————

    My count 2nd time————————————

    My count 3rd time ————————————

    My count 4th time ————————————

    Add total beats and divide by four.

Conclusion:

    My pulse count average for one minute is ————————

    Figure out your pulse rate per:

    ————————1 hour (60 minutes)

    ————————1 day (24 hours)

    ————————1 year (365 days)

If the teacher wishes to pursue this activity further, the students may calculate the averages for girls and boys in the class. This may be plotted on graph paper and can be an exercise in gathering and interpreting scientific information.

BLOOD TYPING CHART

When studying blood students will want to know what blood types can be mingled. A simple blood-typing chart should be made by the teacher on a blackboard or poster for students to copy and interpret.

Materials:

    Rulers, pencils, unlined paper and red pencils.

Procedure:

    1. Have the students print the title and blood types as a heading at the top of their papers.

    2. Draw five horizontal lines about one inch apart and five vertical lines making a total of 16 squares.

    3. At the top of the rows and starting from the left, place the letters of the blood types, A, B, AB, and O. These indicate the blood types of recipients of blood transfusions.

4. At the left of the rows of squares, print the letters A, B, AB, and O indicating blood types of donors.
5. With the red pencils draw a drop of blood within each of the 16 squares. Have the students make dotted the blood that is coagulated and colored in the blood that is clear (figure 4-3).

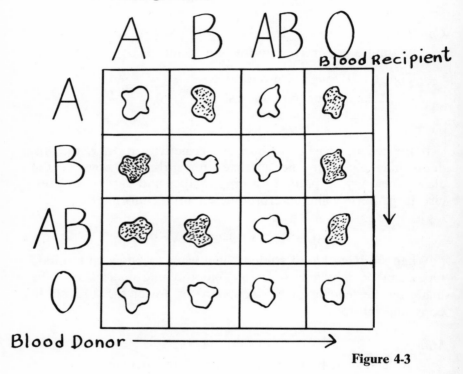

Figure 4-3

The clear indicates blood types that can mix, the coagulated those that cannot.

Interpretation of blood type chart

If you look at blood type A (donor) to the left of the chart and read across, you see that A can give to A and AB. The drops of blood are clear. Blood type A cannot give to B or O. The drops of blood are dotted or coagulated. Blood type O can give blood to any type and is referred to as the universal donor. Blood type O, however, can only receive blood from its own blood type. Blood type AB can receive blood from any type and is referred to as the Universal Recipient. Blood type AB can only give to its own blood type. Students may make up problems about giving or receiving blood such as, "If I was blood type B could I give to blood type AB, or if I was blood type B and needed blood, could blood type A give it to me?"

TEACHER'S BLOOD TYPE

It is not advisable to try to test students for their blood types at this grade level. But they can learn how blood typing is done by a teacher demonstrating by using his own blood. The materials needed are contained in a blood-typing kit, which can be purchased through most biological supply houses.

Materials:

Blood-typing kit, blood lancet or pin, rubbing alcohol, cotton pad, glass slides, overhead projector and colored pencil.

Procedure:

1. The students should write this demonstration as a problem-solving procedure with the aid of the teacher. Since the teacher is using his own blood, it would be natural that the students state the problem, "What is Mr. Smith's blood type?"

2. The procedure is then stated. The teacher will explain that he will place a drop of anti-A serum from the blood-typing kit on the left of the slide and a drop of anti-B serum on the right side.

3. Before the teacher places a drop of his blood into serums A and B, the students should be given a microscope glass slide. They should then trace the slide four times on their paper. With red pencils they may draw in the four outcomes that might take place between the blood and the serum to indicate which of the four blood types the teacher has. A chart similar to figure 4-4 might be prepared by the teacher a few days before the blood-typing demonstration. The clear or coagulated blood drops represented on the slides may be drawn in with red pencils.

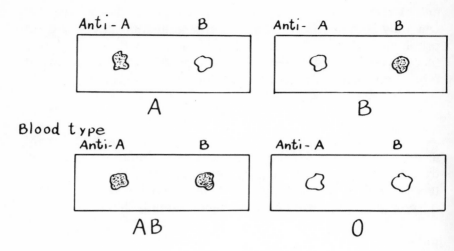

Figure 4-4

4. The teacher may then place anti-A serum on the left of the slide and anti-B serum on the right.

5. He then rubs some alcohol on the cotton pad and rubs it on one of his fingers to sterilize the area. A blood lancet or pin may be used to puncture the skin on a finger for a few drops of blood. Then he would place one drop of blood into anti-A serum and one drop into anti-B serum on the microscope slide. The slide may be placed on an overhead projector and projected onto a screen or wall where the students may view it. They would then compare the slide with the four possible blood types they have drawn on their papers to determine the teacher's blood type. An answer may be written to the problem that was stated.

MAKING A BLOOD SMEAR

Making a blood smear is a way to observe red and white corpuscles.

Materials:

Microscope, microscope slides and Wright's stain.

Procedure:

1. Place a drop of blood about ½ inch from the end of a clean glass slide.
2. Use a second slide to spread the blood thinly on the slide. Position the second slide at an angle above the drop of blood. Touch the upper slide against the blood until the blood is spread along the edge of the slide evenly.
3. Push the upper slide across the bottom slide (figure 4-5). This will make an even smear on the slide.

Figure 4-5

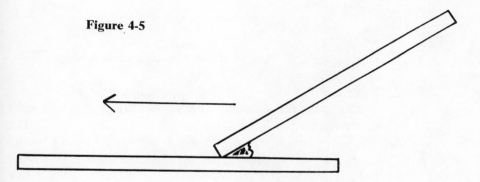

4. Air dry the smear. Then place a few drops of Wright's stain on the smear. Allow to dry. Cover the smear with a glass slide. The techniques of making blood smears might require making several practice slide smears in order to have a good slide. Examine the slide under the high power lens of the microscope. Wright's stain will bring out the nucleus of the white cells and the red cells will appear as pink discs. Have the students make large drawings of the red and white cells.

PILL BOTTLE MODEL OF THE LUNG

The mechanics of breathing in most cases is difficult to show because of the amount of materials involved. Therefore, it is generally demonstrated by the teacher. The pill bottle model of the lung, however, is a simple device that can be built by individual students to demonstrate the mechanics of respiration.

Materials:

Pill bottle with cap, two small balloons, plastic straw and masking tape.

Ask students to bring a pill bottle with snap-on cap, and with the plastic bottom removed. The bottle will be similar to a plastic cylinder with an opening at both ends. Then the mechanics of building the pill bottle model of the lung may be done in the classroom.

Procedure:

1. Cut a piece of plastic straw about 2 inches long. The straw will represent the windpipe.
2. Take one of the small balloons and cut it about 1¼ inches from the bottom. The small end of the balloon will represent a lung.
3. Take the 2-inch straw and place about 3/4 of an inch of it into the balloon. Wrap the excess of balloon tightly around the straw.
4. Take a piece of masking tape and tape the balloon that is twisted around the straw.
5. Now make a hole in the center of the plastic cap. Make the hole large enough so the balloon will fit through it.
6. Take the straw with the balloon attached and pass it through from the underside of the plastic cap so that about 3/4 of an inch of the straw sticks out above the plastic cap (figure 4-6A).
7. Then snap the cap with the straw and balloon in place on top of the pill bottle. The pill bottle represents the chest cavity.
8. Take another small balloon, stretch it and place it on the open end of the pill bottle (figure 4-6B). This balloon represents the diaphragm muscle.

When the balloon outside (diaphragm) is pulled down, the balloon inside (lung) inflates. When the balloon outside (diaphragm) is pushed up, the other inside (lung) deflates. The difference between the inflated and deflated lung is the result of change of air pressure brought about by diaphragm movement.

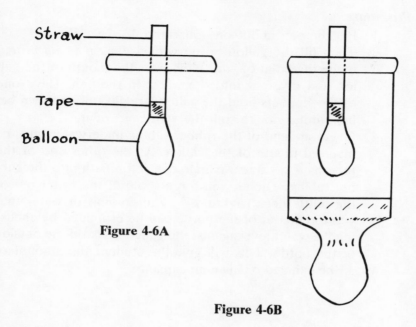

Figure 4-6A

Figure 4-6B

MEASURING LUNG CAPACITY

An instrument for measuring lung capacity is called a spirometer. The type of spirometer that you will construct and use will depend upon the availability of materials in your classroom and the grade level of your students. A description of two different types of spirometers follows. We will call them spirometers I and II.

Spirometer I

Materials:

One gallon bottle, pan, 3 to 4 feet of rubber tubing, empty quart-size milk carton, and tape.

It is recommended that you make a measuring gauge by taking an empty quart-size milk carton, filling it with water and pouring the contents into the gallon bottle. Mark the level of water on the outside of the bottle every ½ quart and quart. Continue filling the gallon bottle with the quart carton up to the four quart mark. This way you will have a measuring gauge of the level of water on the outside of the gallon.

Procedure:
1. Fill the pan about one quarter full of water.
2. Then fill the gallon bottle with water up to its brim.
3. Place the palm of your hand over the mouth of the gallon and invert it into the water in the pan. Have one of the students hold the gallon straight upside down being careful not to spill the water out of it.
4. Place one end of the rubber tubing underwater into the inverted mouth of the gallon. At the other end of the tube have a student exhale one full breath into the rubber tubing. The air will force some of the water out of the bottle neck (figure 4-7). The amount of water that was forced out of the bottle can be estimated by checking the water level against the gauge mark on the outside of the bottle. This will give the student the amount of air he exhaled or his lung capacity.

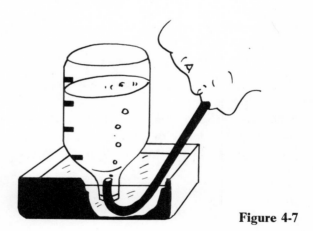

Figure 4-7

Spirometer II

Materials:

One gallon bottle, glass tubing, no. 9 rubber two-hole stopper, rubber tubing, graduated cylinder and an empty quart-size milk carton.

Procedure:

1. Cut a piece of glass tubing about 13 inches long. Make a right angle bend leaving about two or three inches beyond the bend.
2. Take a second piece of glass tubing and bend it at right angle as in step one.
3. Place both pieces of glass tubing through the two-hole rubber stopper opposite each other with their angles about two inches above the stopper.
4. Place a piece of rubber tubing about three inches long over the end of one of the pieces of glass tubing. Place another piece of rubber tubing about seven inches long over the end of the second piece of glass tubing.
5. Fill the gallon bottle about three-quarters full of water.
6. Place the rubber stopper with the glass tubing in the mouth of the bottle.
7. Place the longer rubber tubing into a graduated cylinder. If a graduated cylinder is not available, use an empty milk carton (figure 4-8).
8. Have a student take in a breath and exhale into the short end of the rubber tubing. The amount of water displaced will come out of the opposite glass tube into the carton of milk or graduated cylinder. If the latter, lung capacity can be measured in milliliters.

Figure 4-8

TESTING EXHALED AIR

This is a method for students to test their exhaled air.

Materials:

Test tubes, straws and lime water.

Lime water may be purchased at any local drugstore or may be made by referring to chapter 3, under the heading "How to Make Lime Water."

Procedure:

1. Fill one test tube ¼ full of lime water.
2. Fill a second test tube ¼ full of water. This tube will serve as a control.
3. Have the students exhale through a straw into the plain water for about 30 seconds.
4. Repeat the above procedure into the test tube with the lime water. Let the students compare the two and record the results.

The students will find that the lime water turns milky while the plain water stays clear. This indicates the presence of carbon dioxide gas in the exhaled air. Carbon dioxide gas is one of the wastes caused by the oxidation of burning of foods.

CONVEX LENS AND THE EYE

This activity will give students an opportunity to see how a convex lens is similar to the eye lens and how it focuses an image.

Materials:

Hand lens, 3 × 5 inch index card, yardstick and a pin.

Procedure:

1. Pin an index card, which represents a screen or retina of the eye, at the end of a yardstick.
2. Point the yardstick at the window or an object outside.
3. Move the hand lens along the yardstick until you get a sharp image on the card or "screen." Have the students

describe the image. Is it large or small? Is the object right side up or upside down?

4. Measure the Di (Distance) between the lens and the "screen." This is referred to as the focus or focal point (figure 4-9). The following questions may be asked on the student worksheet: How does this compare with the distance in the eyeball? Should the lens in your eye be thicker or thinner than the hand lens? The distance Di cannot be changed because in the eyeball the distance from the lens to the retina is fixed. How does the eye accommodate the difference?

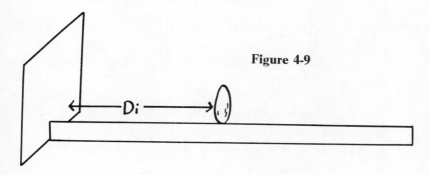

Figure 4-9

OBSERVATION OF A COW'S EYE

Cows' eyes may be ordered from a local butcher shop or a meat packing plant. Packaging and freezing them for storage is best. When the class is ready to observe and dissect them, it is just a matter of defrosting them the day before. By freezing the eyes, the lenses remain clear. The students should be familiar with the external and internal parts of the eye before dissecting them. It is suggested that in preparation the students make a diagram of the eye and label the parts. Below are the materials and a sample worksheet for the observation and the dissection of the eye.

Materials:

Cow's eye, dissecting scissors, forceps and dissecting pans or sheets of newspaper.

Procedure:

1. Examine the gross structure of a cow's eye.
   a. Can you find the muscles attached to the eyeball? What is the function of these muscles?
   b. Look for and identify the sclera, cornea, iris, pupil and optic nerve.
2. Dissect.
   a. Cut through the middle of the eyeball from side to side, not front to back. This will separate the front of the eyeball from the back of it. The optic nerve will be visible on the inside of the back of the eyeball.
   b. Lift off the back of the eyeball.
   c. Carefully remove some of the jelly-like substance that fills the center cavity of the eyeball. Slowly and gently continue to remove the vitreous humor; be careful not to miss the lens which is a round, thick jelly-like mass. Observe the iris, pupil and cornea.
3. Observe and Record.
   Identifying the following (be able to point them out to the teacher):

   Sclera, Color _____ Vitreous humor, Color _____
   Choroid, Color _____ Aqueous humor, Color _____
   Retina, Color _____
   Lens: Description
   Iris: Description
   Pupil: Description
   Optic Nerve: Description

Take the lens out of the eye, wash and clean it. Place the lens over a printed letter and see if it magnifies.

MAKING AN EYE CHART

When studying the eye, students will want to test their vision. If an eye chart is not available, students may make their own eye charts for testing one another.

Materials:

Blank sheet of paper, ruler, black marking pencil or black crayon.

Procedure:

1. Have the students draw seven different printed capital letters 3/8 inches high and 1/16th of an inch thick in the center of their papers. The letters should be about ½ inch apart.
2. Have them darken in the letters with a black crayon or marking pen.
3. They may test one anothers' vision by placing the chart up against the wall about five feet off the ground.
4. Ask them to read the chart from a distance of 20 feet. If a student can read the letters at 20 feet, he has what is called 20/20 or normal vision. Each eye should be tested separately, by covering the eye with the palm and reading. The charts should be changed frequently so the students will not memorize the letter sequence. By having students make their own charts, you can change the sequence frequently.

FINDING THE BLIND SPOT IN YOUR EYE

Students are always amazed that there is a blind spot or a place where there is no nerve ending to receive the light in each eye. This blind spot is a place where the optic nerve enters the eyeball. It can be found through the activity below.

Materials:

Index cards, ruler, and scissors.

Procedure:

1. Measure and cut an index card 1½ × 5 inches long.
2. In the center of the card make a circle and a cross about 3 inches apart (figure 4-10).

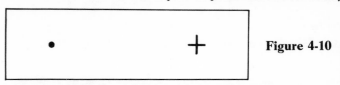

Figure 4-10

3. Close the left eye and with the right hand hold the card at arm's length in front of the right eye.
4. With right eye staring at the circle, slowly bring the card towards your face. There will be a point where the cross will disappear. This is at the exact spot where the cross image falls on the blind spot (or where the optic nerve enters the eye). Reverse the process and find the blind spot in the left eye.

EACH EYE SEES A SEPARATE IMAGE

Each eye produces separate images and the brain combines them into one. These images sometimes overlap to produce a third image between the two.

Materials:

Two pencils.

Procedure:

1. Hold two pencils back to back at the erasers in front of both eyes at arm's length.
2. Stare at the point where the two pencils come together. Slowly bring the pencils towards your eyes and you will see a small third pencil with erasers at both ends in between the other two pencils (figure 4-11). This is the third image produced by the overlapping single images of each eye. The students should try other objects to see if the same effects are produced.

Figure 4-11

AFTER-IMAGE OF THE EYE

When you pass an object rapidly in front of the eye, the retina will retain its image for a short time. If another picture is flashed in front of the eye before the previous image has had time to disappear from the retina, your eye will see both images at the same time. This is referred to as the after-image of the eye.

Materials:

    3 × 5 inch plain index cards, straws, scissors and ruler.

Procedure:

1. Measure and cut a 2-inch square from an index card.
2. Draw a small fish in the center of the square.
3. On the opposite side of the card, draw a fish bowl with water in it.
4. Take a straw and with a pair of scissors slit it about ½ inch from the end.
5. Insert the card with the picture into the slit and push down until it is firmly in place (figure 4-12).

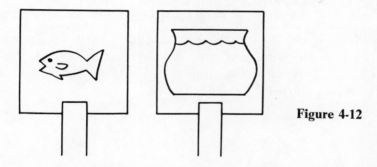

Figure 4-12

6. Place the straw between the palms. Role the straw back and forth.
7. Look at the pictures of the fish and the bowl as they twirl back and forth. It will give the illusion that the fish is in the bowl. Have the students design other combinations.

COMPOSITION OF BONE

The bones of the body provide it with support and protection. As bone cells grow, they deposit minerals such as calcium and phosphorus between them. These minerals make bones hard enough to support the body. This activity is to show that these minerals make bones hard.

Materials:
     Chicken bones, vinegar and a jar.

Procedure:
1. Place several clean chicken bones in a jar and fill it with enough vinegar to cover them.
2. Leave the bones in the vinegar for 4 to 5 days.
3. Remove the bones and rinse them off with water.

Pass them around the class for the students to examine. They will notice that the bones are easy to bend. The vinegar, an acid, has dissolved the calcium and phosphorus compounds that give bones their strength for supporting the body.

MAPPING YOUR FAMILY INHERITANCE FOR EAR LOBES

In humans one of the outstanding inherited characteristics is lobed and non-lobed ears. A lobed ear is one that has a flap of skin hanging down (figure 4-13A). A non-lobed ear has none (figure 4-13B).

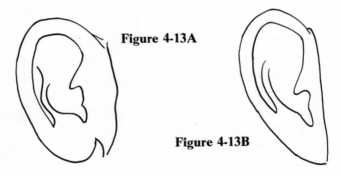

Figure 4-13A

Figure 4-13B

A larger percentage of the population has lobed ears and it is therefore referred to as a dominant characteristic in humans. The non-lobed ear is less common and referred to as a recessive trait. Students may trace and map their dominant or recessive lobes. Have each student go home and observe every member of the family including their mother and father and then return with the information.

Materials:

Clear sheet of paper, ruler, pencils and compasses.

Procedure:

1. Before students map the results, the teacher will have to devise a uniform key of symbols for the class. All males will be represented by a square and all females by a circle. Any person with lobed ears will have his symbol shaded in with pencil and the non-lobed will not.
2. Start with mother and father at the top of the paper, with their children beneath. Attach connecting lines to show family relationships.
3. Place names of brothers and sisters under each symbol (figure 4-14).

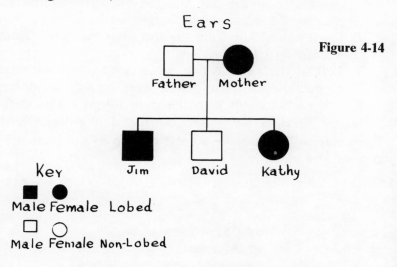

Other dominant recessive characteristics that can be mapped out as above are: brown eyes over non-brown, dark hair over light.

Some students may wish to take a survey of the class to see if lobed ears are dominant.

HUMAN VARIABILITY

This activity is to show students how they inherit differences or variations. A list of inherited characteristics may be placed on the blackboard by the teacher. The list may include the following: brown or non-brown eyes, lobed or non-lobed ears, ability to roll tongue inward or not, hair on back of fingers or not, and right handed or left handed.

The number of students variations may be simply calculated by a show of hands. This information may be totaled on the blackboard and used to make a Human Variability Chart.

Materials:

¼-inch graph paper, rulers, and pencils.

Procedure:

1. Take the ¼-inch paper and count up nine squares from the bottom and draw a horizontal line. Count 15 squares from the left and draw a vertical line so it intersects the horizontal line.
2. On the third line from the top print the title, "Human Variability Chart."
3. Left of the verticle line, list from top to bottom the characteristics from brown eyes to left handedness.
4. For each student with the trait, assign a square from left to right below the horizontal line (figure 4-15).

Each class will then have a total picture of the human variability of the class.

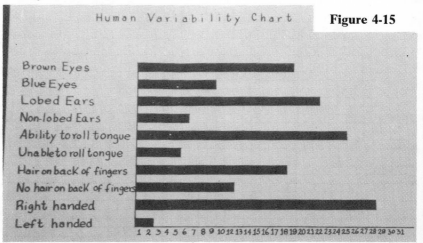

Figure 4-15

— 5 —

ACTIVITIES RELATING
TO PLANT LIFE

Plants can be interesting and exciting when students are involved in observation and exploration of their life functions. The following experiments and learning activities are designed to achieve these ends.

The beginning of the chapter introduces the basic structure of the plant cell. With a tool of science, such as the microscope, the student investigates the inner structures of the cell. The organs of a typical green plant such as the root, stem, flower and leaf (with its food-making process — photosynthesis) reveal themselves through a sequential process of activities and experiments.

The latter half of the chapter deals with various methods of germinating seeds and some of their responses or tropisms. The very end of the chapter deals with the culturing of some non-green plants (fungi) and how to establish different kinds of jar terrariums for the classroom.

LOOKING AT PLANT CELLS

This activity will give students an opportunity to look at two different plant cells.

Materials:

Microscope, slide, cover slide, iodine, toothpick, forceps, small onion, and a leaf of an elodea plant (water plant).

Procedure:

1. Remove the thick peel of an onion and with a pair of forceps peel off a small piece of onion skin.
2. Place a drop of water in the center of a slide and mount the skin into it.
3. Now add a small amount of iodine to stain the skin and place a cover slide over it.
4. Examine the slide under low and high power of the microscope.
5. Have the students make a drawing of several onion cells and label their parts: cell wall, nucleus (small circular body in the cell) (figure 5-1).

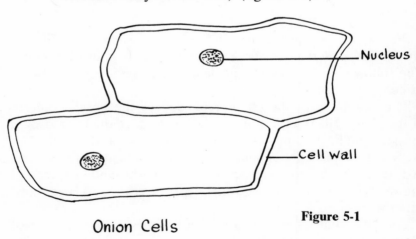

Onion Cells

Figure 5-1

Make a slide of the elodea leaf without staining it. Have the students make a drawing of the cell and compare it with the onion cells. They should notice chloroplasts (green bodies) that contain chlorophyll.

LOOKING AT THE BREATHING
PORES OF LEAVES

This is an activity to see the pores or stomata where the gases of a plant are taken in and given off.

Materials:

Microscope, microscope slide and cover slide, fresh green leaf and forceps.

Procedure:

1. Bend and crack a fresh leaf. With a pair of forceps peel a small piece of the underskin of the leaf.
2. Place a drop of water in the center of the slide.
3. Place the underskin into the drop and cover it with a slide.
4. Examine the openings or stomata under low and high power lenses of the microscope.
5. Have the students count the number of openings they can see and make a large drawing and label the following: stoma (one of the openings), guard cells (two lip-like cells that regulate the opening), and epidermal cells (skin cells) (figure 5-2).

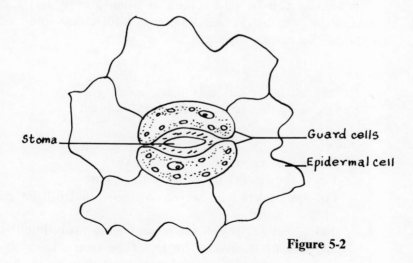

Figure 5-2

HOW YOU CAN SEE THE WATER-CARRYING
TUBES OF A PLANT

The water-carrying tubes or veins can be seen carrying liquids through the stem.

Materials:

Stalk of celery with leaves, beaker, blue or red ink.

Procedure:

1. Fill a beaker one-half full of water and color it with some blue ink.
2. Take a piece of celery with leaves on it and cut off about one inch of the bottom of the stalk.
3. Place the celery stalk in the beaker.
4. Observe how long it takes the ink to travel up the stem to the leaves.

When the ink has reached the leaves, have the students pick the stem apart and trace the veins to the leaves.

HOW LEAVES GIVE OFF OXYGEN

This activity can be undertaken in small groups and also as a short-term demonstration. It permits you to see and collect oxygen as the plant releases it under water.

Materials:

Beaker or large jar, test tube, small glass funnel, small bunch of elodea or any other water plant.

Procedure:

1. Place the water plant in a beaker.
2. Place the funnel over the water plant.
3. Fill the beaker with water. Water should fill the inside of the funnel.
4. Fill a test tube with water and with your thumb over the end of it, invert the test tube over the open end of the funnel. Be careful not to let any air in the test

tube while inverting. Sometimes it takes several at-
tempts before this can be accomplished (figure 5-3).

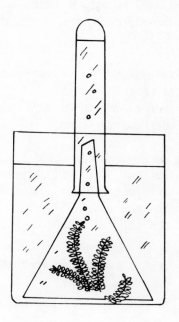

Figure 5-3

5. Place the apparatus in as much direct sunlight as
   possible for four or five days.
6. As a control, repeat the above experiment but place
   the apparatus in a closet for four or five days.

Students will notice that the plant in the sunlight will give
off bubbles of gas which will rise and collect at the end of the
test tube. They can compare the level of gas produced with
that of the control apparatus in the closet. They will then
realize that sunlight is necessary in the production of the gas.

CARBON DIOXIDE ABSORBED BY PLANTS

While studying plants students always read how plants absorb carbon dioxide in the process of manufacturing food
(photosynthesis) in their green leaves. This investigation will
demonstrate its absorption by plants.

Materials:

Two test tubes, two sprigs of elodea (water plant) two to three inches long, two straws, bromthymol blue solution, two corks or rubber stoppers, and a test tube rack.

Procedure:

1. Fill a test tube one-half full of water and add 15 drops of bromthymol blue.
2. Take a straw and blow into the test tube. The exhaled breath containing carbon dioxide will turn the solution yellow.
3. Place a sprig of elodea plant in the test tube and put a stopper at the end of the tube (figure 5-4A).
4. As a control set up another test tube with similar conditions minus the elodea. Place a stopper on the second test tube (figure 5-4B).

Figure 5-4A   **Figure 5-4B**

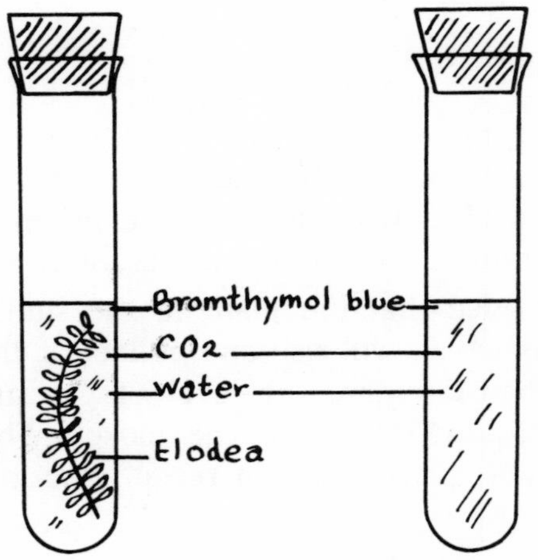

Bromthymol blue
CO_2
water
Elodea

Have the students observe and record the color changes in the two test tubes. Have them check among themselves to see if they had similar results.

PLANTS GIVE OFF WATER THROUGH
THEIR LEAVES

Water loss from the underside of a pore (stomata) of a green leaf is referred to as transpiration. This activity will give students an opportunity to observe this phenomenon.

Materials:

Two small jars (such as baby food jars), a leaf with its stem or petiole, 3 × 5 inch index card, and petroleum jelly or wax.

Procedure:

1. Take an index card and cut from it a three-inch square and punch a small hole in the center.
2. Take a leaf with its stem or petiole attached and place it through the hole in the index card.
3. Fill a jar about three-quarters full of water and place the card on top of it. Make sure the petiole is immersed in the water.
4. Seal the hole around the petiole with petroleum jelly or wax.
5. Cover the leaf with another jar (figure 5-5).

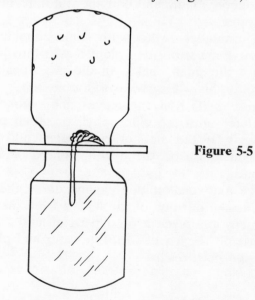

Figure 5-5

After several hours the students will observe moisture collecting inside the jar covering the leaf.

CHLOROPHYLL NEEDED BY GREEN
LEAVES FOR MAKING STARCH

In the first part of this activity, the apparatus is set up in three separate stations. Because of safety measures in heating alcohol, the teacher's supervision is needed. The second portion of this activity is carried on individually or by small groups.

Materials:

Variegated plants (partly green and partly white) such as ivy, coleus or geranium, isopropyl alcohol (rubbing), iodine solution, forceps, eyedroppers, petri dishes, two different size beakers (one to fit inside the other), alcohol lamp, wire gauze, ring clamp with stand or a hot plate heating unit, and safety glass if available.

Procedure:

1. Set up about three stations with an alcohol lamp with ring clamp attached to a stand with a wire gauze. Light the alcohol lamp and boil about a third of a beaker full of water.
2. Fill a smaller beaker with some rubbing alcohol and place some variegated plant leaves into it.
3. Place the small beaker in the large beaker of boiling water. This will serve as a double boiler or water bath (figure 5-6). Boil the leaves and alcohol for several minutes until the chlorophyll or green pigment is removed. If the alcohol evaporates, add more to the beaker. Turn off the heating source or extinguish the flame.
4. Have individual students come up to the stations and each take out one of the leaves from the alcohol with forceps and place it on a petri dish.
5. Rinse off the leaves with tap water and place them flat on the petri dishes.

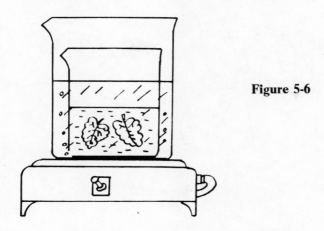

Figure 5-6

6. Place some drops of iodine solution on the leaves. The part of the leaf that had chlorophyll changes to blue or blue-black, showing the presence of starch, while the white part of the leaf (no chlorophyll) has no change in color. This shows that chlorophyll is needed in the process of making food or photosynthesis in green plants.

Sunlight is needed to make starch in green leaves

The materials for testing the starch in leaves are basically the same as in the previous activity except for the following.

Materials:

Green plants with solid green leaves, corks and pins or black paper with paper clips.

Procedure:

1. Cut the corks into disks and pin them on both sides of a green leaf (figure 5-7). Place several on each leaf. If corks are not available, use black paper or tin foil. Attach it to the leaves with paper clips.

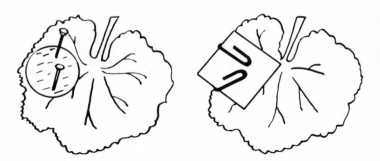

Figure 5-7

2. Place them in the sunlight for several days.
3. Have the students pick the leaves from the plants, remove the disks and test the leaves for starch as in the previous activity.

The difference in color between the covered spots and the rest of the leaf demonstrates that sunlight is necessary for making food in green plants.

HOW GREEN PLANTS REACT TO SUNLIGHT

This investigation allows students to discover how a plant will react to sunlight. Students may plant seeds and germinate their own plants for this investigation.

Materials:
    One lima bean, 2 empty quart-size milk cartons and soil.

Procedure:
1. Cut down a quart-size milk carton to about 3½ inches and fill it three-quarters full of soil.
2. Plant a lima bean ½ to 1 inch into the soil in the center of the carton and water the soil.
3. When the bean begins to sprout, cut off the end of the second quart carton. On one of the sides near the top, cut out a one square-inch hole.

4. Place this carton on top of the one with the seedling (figure 5-8).

Figure 5-8

5. Make the hole face the window.
6. Keep the seedling watered. As a control set up another seedling apparatus at the same time, but do not cut a hole. Leave both for about 10 days after the seedling sprouts. At the end of this time have the students uncover both plants and compare the growth and the leaves.

HOW MUCH WATER A PLANT USES

This is a method of measuring the amount of water used by a plant.

Materials:

Cutting from a philodendron, 3 × 5 inch card, flex-straw, small piece of clay about the size of a pea, Scotch Tape and paper clip.

Procedure:

1. Fold the 3 × 5 card in half and then fold the edges back so that the card is in quarters.
2. Open the edges so the card will appear as a "T" facing you. Tape the top and bottom of the card with Scotch Tape (figure 5-9A).
3. Take the flex-straw and bend it into a "J" at the flexible end and tape it to the face of the card so that both ends of the straw stick above the card (figure 5-9B). Place a paper clip on the back bottom of the "T" for balance so the card will not fall forward.

Figure 5-9A

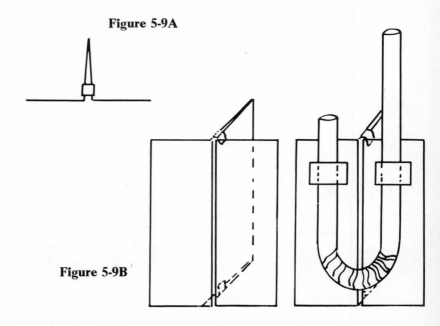

Figure 5-9B

4. Take a cutting from a philodendron, place it in the short end of the straw and fill with water from the opposite end. Use clay to seal the space around the plant to prevent evaporation (figure 5-10).

The level should be marked and recorded daily in order to know how much water the plant uses in 24 hours.

As a control another device could be set up with similar con-

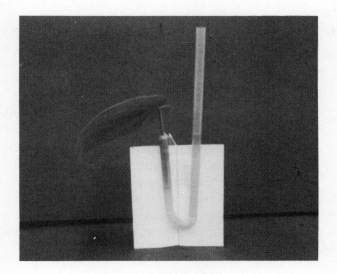

Figure 5-10

ditions minus the plant. One end of the straw should be sealed with clay. This control should tell you how much water is evaporated by air compared to the amount used by the plant.

OBSERVING PARTS OF A FLOWER

In order to learn the structures of a flower, it is easier for the students at this grade level to learn through a diagram or model before dissecting a flower. This prevents confusion among the students when dissecting and trying to identify the parts of a flower. It will also be found that if all the students have similar flowers it is helpful in identifying and comparing the flower structures. You will find that contacting local plant nurseries or florists will help obtain similar flowers in quantity. Many of these establishments would be willing to donate flowers for your classes for this worthwhile activity. A worksheet may be prepared by the teacher with such questions as below.

Materials:
    Flowers, forceps, scalpels and hand lenses.

Procedure:

1. Count the number of petals on the flower you are examining. If there are more than 10, just estimate the number.
2. How many pistils are there? If there are too many to count, just estimate the number.
3. How many stamens are there? If there are too many to count, just estimate the number.
4. Gently pull off some of the petals on one side of the flower to expose its center.
5. Cut the flower lengthwise to expose the ovary with its ovules. Notice their arrangement.
6. Make a drawing and label the following parts: petal, sepal, pistil, stigma, style, ovary, ovule, stamen, filament and anther (figure 5-11).

Figure 5-11

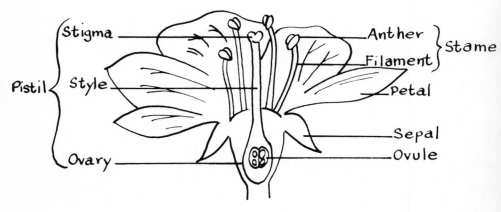

If microscopes are available, have the students examine some of the pollen grains from the anther and draw them.

STUDYING A TYPICAL TAPROOT

The function of the roots is to absorb minerals and water. Roots such as carrots and beets are used as a source of food. This exercise is to have the students observe, dissect, draw and label the parts of a typical taproot.

Materials:

Carrots, scalpels or knives, and unlined paper.

Procedure:

1. Hand out unlined paper, scalpels and carrots to the students. Have them place a title such as "General Structure of the Taproot."
2. Have them examine and draw the external view of the carrot. Label the primary and secondary roots (figure 5-12A).
3. Cut the taproot longitudinally through its center. Examine and draw the longitudinal section. Label the cortex (storage area), central cylinder (where conduction occurs), and secondary roots (small branch roots) (figure 5-12B).
4. Cut a cross section of the carrot and make a drawing of it. Label the structures you can identify that are similar to the longitudinal section (figure 5-12C).

Figure 5-12A   **Figure 5-12B**

Figure 5-12C

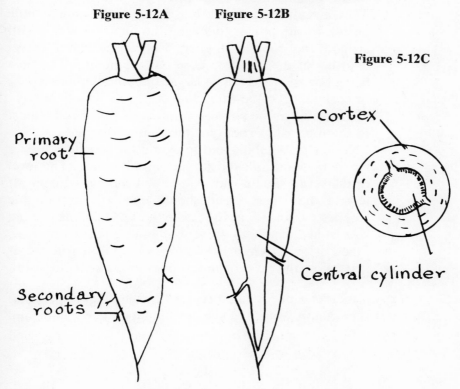

Primary root

Secondary roots

Cortex

Central cylinder

BUILDING AN OSMOMETER

This is an instrument that will measure the rate of a liquid passing through a living plant membrane, illustrating the principle of osmosis. This activity is done in small groups.

Materials:

A large carrot or potato, one-hole rubber stopper or cork, 1 piece of glass tubing that fits the one-hole rubber stopper, scalpel, cork or apple borer, molasses or syrup, a candle, and a beaker or bowl.

Procedure:

1. With a knife or scalpel cut off the end of a potato so it will stand upright without falling. Peel part of the lower end of the potato.
2. Take the one-hole rubber stopper or cork and place it on top of the potato and trace the circular outline with a pencil onto the potato.
3. Take a scalpel or a cork borer and cut a circular hole on top of the potato following the line that was traced around the rubber stopper. Cut down about three-quarters of the way and bore out the potato. The hole being bored should not be larger than the rubber stopper or cork that is to be fitted into it.
4. Fill the hole in the potato with molasses or syrup almost to the top. Allow room for the rubber stopper.
5. Take the glass tubing and fit it into the one-hole rubber stopper or cork. Fit the stopper into the hole. The level of the syrup should show slightly in the glass tubing. If any of the excess syrup should ooze out around the stopper or potato, wipe it away and drop some melted wax from a candle to prevent any more leakage.
6. Place the potato in a beaker or bowl. Fill it with water just below the level of the stopper or cork (figure 5-13).

If the osmometer needs some support in standing, use a stand and test tube clamp. Mark the level of the liquid in the tube and

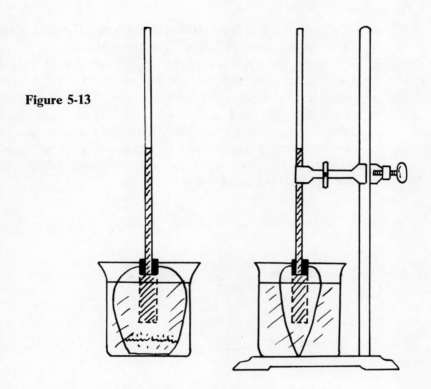

Figure 5-13

keep a record on how fast it climbs. This movement of water through the cells of the potato from a region of greater water concentration to a region of lesser concentration is called osmosis.

EXAMINING A LIMA BEAN SEED

Lima bean seeds are suggested in this activity because they illustrate a dicot seed — one with two seed leaves. This activity is performed individually or in groups of two.

Materials:

Two lima bean seeds (a dry seed and one that has been soaked overnight), scalpel and hand lens.

Procedure:

1. Compare the wet bean with the dry one. How do they differ in appearance? How are they alike?
2. Notice on the concave part of the bean a scar called the hilum. This was the point where the bean was attached to the pod. Take a hand lens and look directly above the scar and you will see a tiny pore or hole called the micropyle, this is the spot where water enters during germination. Make a side-view external drawing of the dry bean. Label the seed coat, hilum and micropyle (figure 5-14).

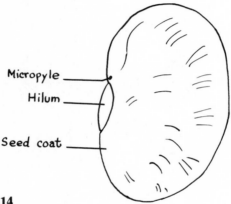

Figure 5-14

3. Take the soaked bean and cut or break through the seed coat and carefully separate the two halves of the bean. These halves are called cotyledons or seed leaves, and they contain stored food. Because they have two seed leaves or cotyledons, beans are classified as dicotyledons.
4. Between the two cotyledons, look for some tiny leaves, the beginning of a stalk, and a tiny root tip. This is the embryo bean plant.
5. Draw an enlarged diagram of an internal bean seed and label the seed coat, plumule (becomes tiny leaves), stem of the plant hypocotyl (becomes the root), and the embryo and cotyledons (figure 5-15).

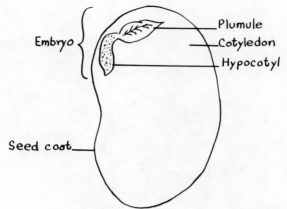

Figure 5-15

METHODS OF GERMINATING SEEDS

Plant greenhouses

This method is similar to greenhouses in that it keeps in the moisture and heat that is needed for better growth and requires less watering in the germination of seeds.

Materials:

One empty quart-size carton, any seeds you wish to germinate, a large plastic bag and soil.

Procedure:

1. Take an empty quart-size milk carton and cut out one of its sides so it will look similar to a box without a cover.
2. Fill the carton three-quarters full of soil.
3. Plant the seeds you wish to germinate about one inch under the soil and saturate the soil with water.
4. Take a large plastic bag and cover the entire carton. Secure the open end with a rubberband or tuck it under the milk carton on all sides and tape it (figure 5-16).

Figure 5-16

Place the greenhouse in the sunlight or other warm spot. In about a week the seeds will germinate. Remove the plastic bag when the seedlings are a few inches high. Then continue to water your plants. They may then be transplanted to individual flower pots.

Beaker germination

Materials:

Beaker, blotting paper, lima beans or any other seeds to be germinated.

Procedure:

1. Cut a piece of blotting paper and line the beaker wall with it.
2. Place some seeds between the blotting paper and the beaker's glass wall (figure 5-17).

Figure 5-17

3. Now fill the bottom of the beaker with water and let the blotting paper become saturated. Continue adding water to the bottom so that the blotter is always saturated. Check the water table daily. To prevent the blotting paper from drying out over the week-end, place in the beaker some crumpled paper towel, cotton or sponge and add water.

Glass plate germination

Materials:

Two small plates of glass, blotting paper, rubberbands, lima beans or any other seeds desired, and a shallow pan.

Procedure:

1. Cut a piece of blotting paper the same size as the glass plate.
2. Wet the blotting paper and lay it on a glass plate.
3. Place a few lima beans on the blotter.
4. Place the other glass plate over the seeds.
5. Hold the two glass plates together and place some rubberbands around them.
6. Stand the plates on edge in about ½ inch of water in a shallow pan (figure 5-18).

Figure 5-18

GERMINATING SEEDS GIVE OFF
CARBON DIOXIDE GAS

This activity will demonstrate that germinating seeds oxidize their stored food and give off carbon dioxide gas.

Materials:

One quart-size jar with lid, 1 dozen lima beans, 1 test tube, lime water and cotton.

Procedure:

1. Place a layer of cotton in the bottom of the quart jar and saturate it with water.
2. Soak about a dozen lima beans in water for several minutes and then place them on top of the cotton.
3. Fill ¼ of a test tube with lime water. Lime water indicates the presence of carbon dioxide gas by changing to a milky color. Lime water may be purchased at a local drugstore or can be made by referring to chapter 3.
4. Place the test tube with the lime water upright in the quart jar. Lean the test tube against the inner wall of the jar so the limewater will not spill. Screw on the cap to the jar (figure 5-19).

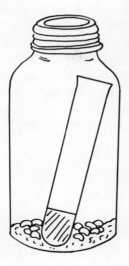

Figure 5-19

Another jar may be set up as a control or a comparison. Set up the control condition the same as above minus the beans. Observe the jars and notice that in time the limewater with the beans will turn milky indicating the beans are giving off carbon dioxide gas.

HOW MUCH PRESSURE A SEEDLING EXERTS

This activity will demonstrate the amount of upward pressure exerted by a growing seedling by having the soil above it covered with paraffin wax.

Materials:

Lima bean seeds, drinking glass, wax paraffin, soil, rubber-bands and rulers.

Procedure:

1. The students should fill a drinking glass ½ full of soil.
2. Have them plant the lima beans about 1 inch deep against the inner wall of the glass so they can be observed from outside. Water the soil and bean.
3. Students should measure with a ruler different distances above the soil surface such as ¼, ½, 3/4, 1 inch,

etc. A rubberband should be placed around the outside of the glass marking the exact height the wax is
poured. The teacher should have some source of heat
and a container to melt some paraffin wax. The amount
needed will depend on how many students are taking
part in the activity. The students should then stand in
line and have the teacher pour the melting wax to the
exact measurement indicated by the rubberband (figure
5-20).

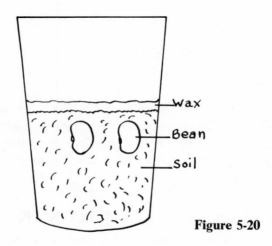

Figure 5-20

The students should then record the date of planting of the
seedling and the thickness of the wax. As the seedlings start to
grow, they will break through the different thicknesses of wax.
In this way the students may determine the amount of pressure
exerted by the seedlings.

HOW ROOTS RESPOND TO GRAVITY

In order to see how roots respond to gravity, some seeds
must be germinated in order to observe their roots. This activity
will show students that roots will grow towards gravity.

Materials:

Two small plates of glass, blotting paper, rubberbands, 4 lima beans and a shallow pan.

Procedure:

1. Cut a piece of blotting paper the same size as the glass plates.
2. Wet the blotting paper and lay it on a glass plate.
3. Place the four lima beans on the blotter.
4. Place the other glass over the seeds.
5. Hold the plates together and place rubberbands around them.
6. Stand the glasses on edge in about ½ inch of water in a shallow pan.

When the roots start to sprout, the students will notice that they turn down. Have students turn the plates so that the roots are in an upright position. In time the roots will turn downward again.

JAR TERRARIUM

This activity is a way of making a miniature garden of soil and plants; or a woodland or desert habitat, from a gallon mayonnaise or mustard jar. These gallon jars are generally obtainable from the school cafeteria.

Materials:

One gallon jar, 2 pieces of wood about 11 inches long, 4½ inches high and 3/4 inches thick.

Groups of students may lay the jars on their side and may plant different kinds of miniature gardens or habitats. Two pieces of scrap wood can be cut to cradle the circular jar and prevent it from rolling (figure 5-21).

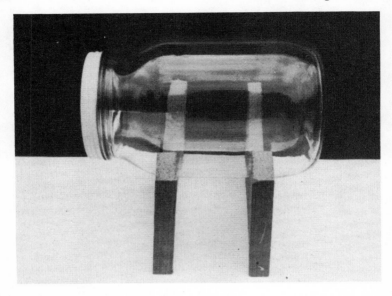

Figure 5-21

Woodland terrarium

Procedure:

1. Place a bottom layer mixture of sand and fine gravel.
2. The second layer from the base should be a mixture of loam and decaying leaves. Moisten soil.
3. Any woodland plants may be used such as mosses, ferns, liverworts. Do not overcrowd the terrarium. When the terrarium has been planted, sprinkle some water around each plant and close the jar. Place the terrarium where there is plenty of light.

Desert terrarium

Procedure:

1. Place a bottom layer of fine gravel.
2. The second layer should be a mixture of sand and soil.

3. Plant miniature cacti or other desert plants or succu-
   lants. Sprinkle water around them after planting.

School ground survey terrarium

This will give an opportunity for students to survey a portion
of the school grounds that is not cultivated.

Procedure:

1. Place a bottom layer of gravel for drainage.
2. The second layer should be the soil from the ground
   you are surveying.
3. Dig up sample grasses or weeds from the area and
   plant a few of them. Water the soil and plants. Cover
   terrarium and observe to see if any other plants sprout
   from the soil.

CULTURING YEAST

The common yeast is a good example of a one-called fungus
that reproduces by dividing in a manner called budding.

Materials:

One cake yeast or a package of dry yeast, 1 test tube, 1
teaspoon of sugar, stirring rod, and a wad of cotton.

Procedure:

1. Fill a test tube ½ full of warm water.
2. Crumble ¼ of a cake of yeast and add it to the water
   in the test tube.
3. Add one teaspoon of sugar and mix the solution
   thoroughly with a stirring rod.
4. Plug the opening of the test tube with a wad of cotton
   and set it in a warm place.

The next day have the students place a few drops of the
solution on a microscope slide and cover it with a cover glass.
Examine the slide under the microscope and notice that the

yeast cells are round and are producing new cells by budding (figure 5-22). Make a sketch of the cells.

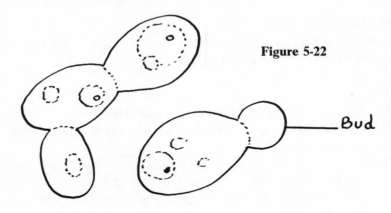

Figure 5-22

Bud

CULTURING BREAD MOLD

This activity will give students a simple way of culturing and observing molds. Molds are a fungi or non-green plants.

Materials:

One slice of bread, petri dish or bowl, blotting paper.

Procedure:

1. Place a piece of blotting paper on the bottom of a petri dish or bowl and wet it with water.
2. Place a piece of sliced bread that has been exposed to the air several days on the wet blotter.
3. Sprinkle some water on the bread, cover the petri dish or bowl and place it in the dark for several days. Check to see that the bread stays moist. After several days a black mold will develop on the bread.

Students may examine a sample of the mold under the microscope. They will be able to see the mold, which is made up of tiny clusters of grayish threads with black knobs at the end called sporangium. The sporangium may be broken open and the spores seen (figure 5-23). The spores will reproduce new mold.

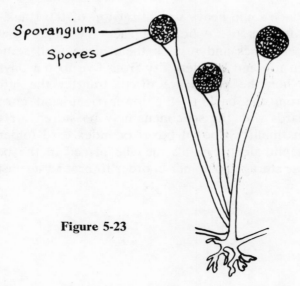

Sporangium

Spores

Figure 5-23

Students may want to investigate to see if mold will grow on the other objects such as shoe leather, cloth and paper.

PREPARING A HERBARIUM

A collection of plants pressed, dried and mounted on paper or cardboard is referred to as a herbarium. The value of this type of collection is that it may aid the student in identifying and comparing plants before going on a field trip. It may last for years and may be used to show certain ecological regions.

Materials:

5 × 8 inch index cards and transparent tape.

Procedure:

1. Place the plant or part of it to be mounted in the center of the index card and tape it down with transparent tape. Place the name of the plant, date collected, and name of the collector on back side with a pencil.
2. Place paper towels over the specimen with a blank index card between every two towels as you build the stack.

3. Place some heavy weight on top of the stack.
4. After about 24 hours replace the paper towels with fresh ones and restack them again with some weight on top and let them dry from five to ten days.
5. When the plants are dried, transfer the information from the back to the lower right-hand corner. The cards with the specimens may be stored for later use in small cardboard boxes or index card folders. Some naphthalene crystals may be placed in the bottom of the storage container in order to protect against insects.

—6—

ACTIVITIES THAT RELATE TO INSECT LIFE

This chapter on insects was developed because students generally bring more insects to the classroom than any other living things. This fascination of collecting insects becomes a motivating factor for activities on how to collect, preserve, store, display, and study insects. The first half of the chapter introduces activities on how to accomplish these ends.

An appreciation of the remarkable changes and behavior of insects can be obtained by students through the culturing and observation of mealworms. The ecological relationship between the oak gall wasp and the growth of the gall on an oak tree with its by-products is observed and investigated in the latter half of the chapter.

MAKING AN INSECT PREPARATION JAR

Whenever students collect insects, there is the problem of killing them. It is wise for students to make insect preparation jars under the supervision of the teacher. Encourage students to collect and bring to class wide-mouth plastic or glass jars of different sizes with caps. The size of the jar will depend upon the kind of insects you wish to collect. It is convenient to have several different sizes. You may not wish to mix butterflies with other insects because they may damage the butterflies' wings.

Materials:

Wide-mouth jar, blotting paper or corrugated cardboard, adhesive or masking tape, and carbon tetrachloride (cleaning fluid). Carbon tetrachloride may be purchased from a drugstore.

Procedure:

1. Place a layer of cotton on the bottom of the jar. Cotton should be charged with carbon tetrachloride before you go on a field trip. Since students have a tendency to saturate the cotton, emphasize to them the importance of using a small amount of liquid. It is the evaporation of the liquid that kills the insect.
2. Cut and puncture a few small holes in a circular piece of blotting paper or corrugated cardboard and place it snugly inside the jar above the cotton. There should be a space between the moist cotton and cardboard.

It is recommended that you place a piece of cleaning tissue on the bottom of the cardboard before collecting insects. The tissue will absorb the moisture given off by the captured insects.

MAKING A TEMPORARY INSECT NET

When going out to capture larger winged insects such as butterflies and moths, it is necessary to have some sort of sweep net in order to capture them without damaging their wings. This activity is a method of showing students how to make their own insect nets in the classroom before going on an outing or field trip to capture insects.

Materials:

One wire clothes hanger, 1 clear 11½ × 13 inch plastic all-purpose storage bag, masking tape, 1 pair of pliers and 1 ruler.

Procedure:

1. Take a pair of pliers and use it to untwist and straighten a metal hanger. Do not straighten the U-shaped end of it (figure 6-1A).
2. Bend the straight end of the wire into a circular ring about 7½ inches in diameter leaving about 1 inch at the end straight. Tape this end to the wire shaft (figure 6-1B). Tape so they will not come apart.
3. With the clear plastic storage bag, overlap the circular rim. Tape the bag to itself over the rim. Place tape at intervals around the rim. The shaft of your net will be a straight piece of wire approximately 16 inches long with a "U" end to serve as a handle (figure 6-1C).

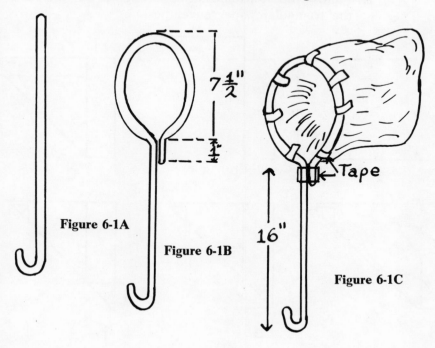

$7\frac{1}{2}''$

Tape

Figure 6-1A

Figure 6-1B

16"

Figure 6-1C

MAKING TRIANGULAR PAPER ENVELOPES

This method is used to temporarily and safely store large dead insects such as moths, butterflies and dragonflies to prevent damage until they can be mounted and stored in a collection. It consists of triangular envelopes that can carry the insect and be sealed. Students can make these envelopes before going on a field trip. I find it practical to have students make two different sizes.

Materials:

Unlined paper, rulers and scissors.

Procedure:

1. Measure and cut out a piece of paper 5 × 7 inches. This larger size is for large butterflies such as swallow tail. The second size 4 × 6. This can be used for smaller butterflies.
2. Follow the steps illustrated in figure 6-2 for making the triangular paper envelopes.

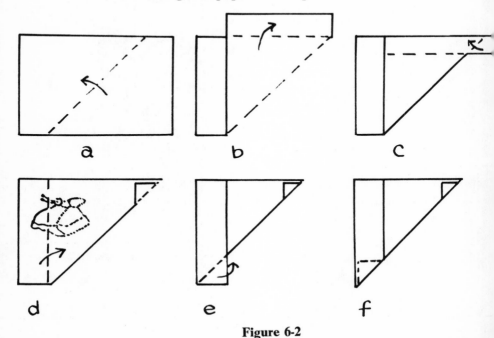

Figure 6-2

3. When an insect is closed in the envelope, data such as location, date and the collector's name should be written on the outside.

MAKING AN INSECT PIN GAUGE

Whenever you pin insects and place their labels, there is the problem of having them all at a uniform height in the collection. This is a simple activity where students can build their own corrugated cardboard insect pin gauge to achieve this uniformity.

Materials:

Small piece of corrugated cardboard, ruler and scissors.

Procedure:

1. Measure and draw a 1 × 3 inch rectangle on a piece of corrugated cardboard.
2. Place the rectangle on its side and measure vertically 1", 3/4" and ½" at 1" intervals from left to right. This will make it look like a staircase (figure 6-3). Cut out the insect gauge.

The top step is the height for pinning the insect. By your holding the pin gauge upright, the pin going through the insect can be placed in one of the spaces in the top of the corrugated cardboard. The second and third steps are for labeling the specimen.

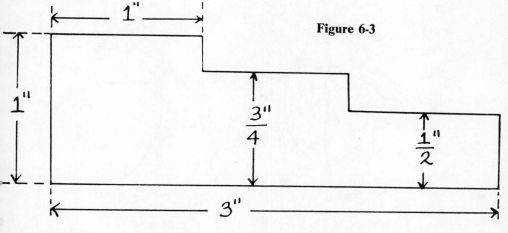

Figure 6-3

PINNING AND MOUNTING INSECTS

The best way to preserve and keep an insect is by pinning it with insect pins. Straight pins are not desirable because they are short and thick and therefore tend to break the body of the insect. Insect pins are longer and range in sizes from 00 to 7. Size 3 is generally appropriate for an average insect. The longer pins leave room beneath the insect for the mounting of labels. Insect pins might be purchased locally or ordered through a science supply house.

Insects such as butterflies and moths are pinned in the center of the thorax (figure 6-4A). Flies, bees, wasps, and grasshoppers should be pinned slightly to the right of the thorax midline (figure 6-4B). Beetles should be pinned near the front right wing (figure 6-4C). Bugs should be pinned slightly right of midline through the triangular scutellum (figure 6-4D).

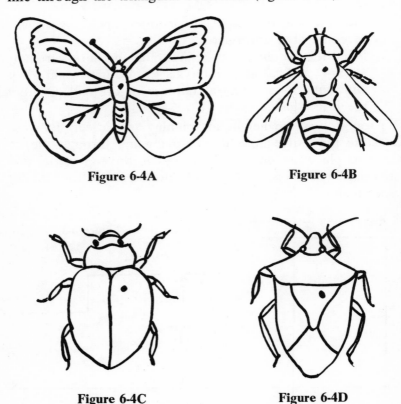

Figure 6-4A   **Figure 6-4B**

Figure 6-4C   **Figure 6-4D**

Materials:

Insect, insect pin, 3 × 5 inch white index cards, scissors, ruler and insect pin gauge.

Procedure:

1. Hold the insect with your left thumb and forefinger and with the right hand insert the pin in the proper spot in the insect.
2. Place the pin through the corrugated cardboard on the first step of the insect pin gauge (figure 6-5A).
3. Take an index card and measure several labels ½ × 1 inches. These labels are slightly larger than recommended by entomologists. I find that students at this grade level have difficulty printing the information on a smaller label.
4. On the first label print the locality where found, date, and the collector's name.
5. From the top step take the pin with the mounted insect, mount it through the center of the label and pass the pin through the second step of the insect gauge (figure 6-5B).
6. If the insect can be identified, print its scientific name and its common name on a second label. Place your pin with the insect and locality label through the center of the identification label and pass it through the third step of the insect gauge (figure 6-5C). The labels should be parallel with the mounted insect.

Insects that are too small to be pinned may be mounted on a triangular card about ½ inch long and 3/8 inches wide at the base. An insect pin is inserted near the base of the triangle and the insect is glued at the tip of it (figure 6-6). Then appropriate labels are placed under the insect as above.

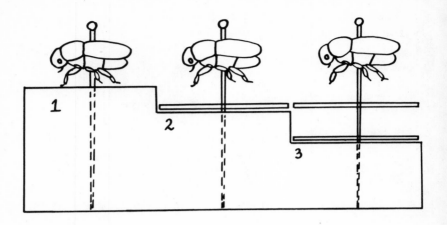

Figure 6-5A   Figure 6-5B   Figure 6-5C

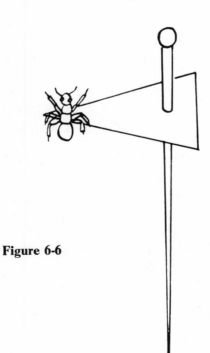

Figure 6-6

PRESERVING THE INSECT COLLECTION

Once the insects are placed in a box, the collection must be protected from being eaten by live insect pests.

Moth ball mount

Naphthalene moth balls are a good insect repellent. By mounting the moth ball, you prevent it from rolling around in the box and knocking over the insect specimens.

Materials:

Mothball, straight pin, candle and forceps.

Procedure:

1. Light a candle and heat in its flame for a few seconds the head of a pin by holding the point with a pair of forceps.
2. When the head of the pin is hot, push it into the center of the mothball. Hold the mothball and pin still until they cool. Then the ball will stick to the pin (figure 6-7). The mothball mount can then be pinned in the bottom corner of the storage boxes. As time passes the mothball will evaporate and another fresh mothball mount will have to be set up.

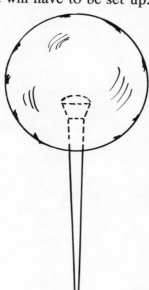

Figure 6-7

Insect repellent holder

For larger collections, students may desire to make an insect repellent holder, which is a small box for holding napthalene flakes that can be pinned in an insect storage box. If a great quantity is to be made, it is easier for the teacher to make the pattern and duplicate it. The students cut out the pattern, trace it on index cards and construct the boxes to be placed in insect collections.

Materials:

> Napthalene flakes, 5 × 8 inch index card, 1 square inch screen mesh, transparent or masking tape and ruler.

Procedure:

1. With a ruler draw a rectangle 2¼ inches long and 1 1/8 inches wide on an index card.
2. Measure 5/8 of an inch from the left side and draw a rectangle the same size but crossing the first rectangle. When completed it will look like a cross (figure 6-8A). This will serve as the bottom of the holder.
3. Draw two more rectangles, 2½ inches long by 1¼ inches wide, and intersect them as in step 2. Measure a 3/4-inch square in the center, where the two rectangles intersect.
4. Cut out the two crosses and also the 3/4-inch square in the center of one (figure 6-8B).
5. Cut a 1 square inch piece of screen mesh and tape the edges of it behind the 3/4-inch square hole in the center of one of the crosses.
6. Fold the four flaps (as shown by the dotted lines) on both crosses and tape the edges together to make two small boxes.
7. Fill the box without the screen window with napthalene flakes.
8. Fit the box with the screen window over the one filled with napthalene (figure 6-9). The insect repellent holder may be pinned to the lower right or left corner.

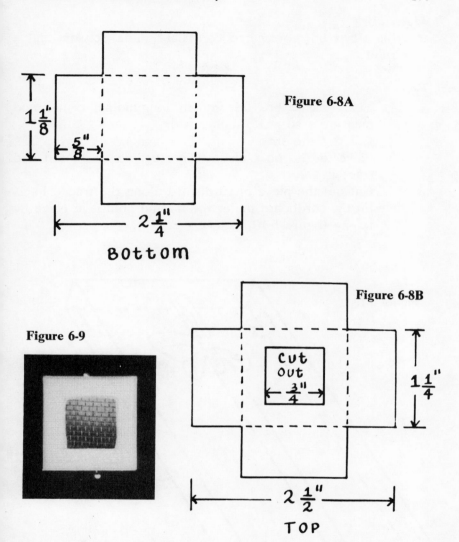

Figure 6-8A

Bottom

Figure 6-9

Figure 6-8B

Cut
Out
$\frac{3''}{4}$

$1\frac{1}{4}''$

$2\frac{1}{2}''$

Top

MAKING A SHOE BOX SPREADING BOARD

A spreading board is a device for pinning and spreading out the wings of an insect to dry. Spreading boards are not always available in the classroom when needed. This activity is a way for students to build their spreading boards.

Materials:

Shoe box lid, corrugated cardboard, tape, scissors and a ruler.

Procedure:

1. Take a shoe box lid, find its longitudinal center and measure a strip 3/8 inches wide.
2. Cut out the strip.
3. Place the lid on a piece of corrugated cardboard and trace it.
4. Cut out the piece of cardboard along the traced lines. Fit the cardboard under the lid and hold it in place by taping (figure 6-10).

Figure 6-10

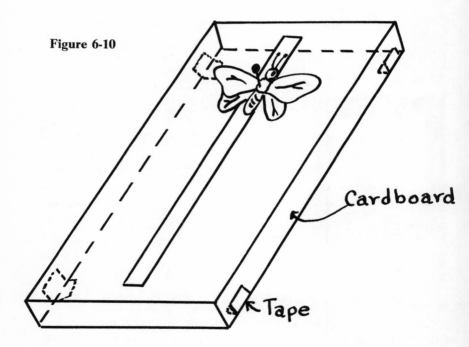

An insect such as a butterfly or moth that has been pinned through the thorax is placed in the groove in the spreading board and the wings are spread out and kept in place by pinning strips of flexible cardboard over them. Drying time is about five to seven days. A good way to check is to touch the abdomen with a pin. If the abdomen is dry and stiff, the specimen may be removed and placed in the collection.

MAKING INSECT STORAGE BOXES

One problem of collecting and pinning insects is the need to store them. Students may make their own homemade storage boxes.

Materials:

> Cigar box or any other small box that is deep enough for pinned insects, corrugated cardboard, sheet of paper and scissors.

Procedure:

1. Take a sheet of paper and trace the bottom of a cigar box.
2. Cut out the paper pattern and trace it on corrugated cardboard.
3. Cut out the cardboard and place it in the bottom of the cigar box (figure 6-11). The cardboard should fit snugly so it will not fall out. If the cardboard is loose, glue it to the bottom. The purpose of the corrugated cardboard is to make a soft bottom that will permit easy pinning of the insects. The storage box may be covered with colored paper or painted to make it look attractive.

Arrange the pinned and labeled insects in rows. Pin a moth ball mount or an insect repellent holder with napthalene flakes in the lower left or right of the storage box for the protection of the collection.

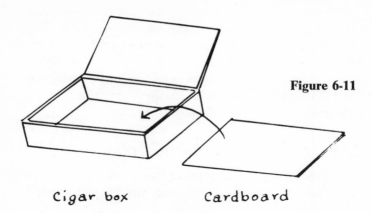

Figure 6-11

Cigar box   Cardboard

MAKING A DISPLAY MOUNT

Students not wishing to pin their insects may display them in a container. These mounts can display one large or many small insects.

Materials:

Small cardboard box (such as for jewelry or handkerchiefs), roll of cotton, transparent sheet of plastic or plastic film, transparent tape, 2 straight pins and scissors.

Procedure:

1. Cut out a square window in a box lid. Leave a boarder around the edges of the window.
2. Turn the lid on its face and with scissors cut a piece of clear plastic or plastic film to fit the window. Use transparent tape to hold the plastic in place against the window edges.
3. Now line the bottom of the box with a layer of cotton.
4. Place specimen or specimens with appropriate labels. Before closing the lid, place several mothballs or Napthalene flakes under the cotton in the corners.
5. Place the lid over the specimens and put two straight pins through opposite sides of the box. They hold the mount in place (figure 6-12).

Figure 6-12

INSECT LIFE CYCLE DISPLAY BOX

At times students preserve and display the life cycle of an insect. Eggs, larvae and pupae are too soft to be pinned and dried. They must be placed in a container with some liquid preservative and then displayed. This activity will demonstrate how students set up the life cycle of an insect for display.

Materials:

Cigar box, small pill bottles with caps or small vials with corks or stoppers, corrugated cardboard, rubbing alcohol (isopropyl), flexible wire, colored paper, ruler and scissors.

Procedure:

1. Take each stage of the insect life cycle you wish to display and place it in a vial. Fill each vial three-quarters full of rubbing alcohol and seal with cap or stopper.
2. Cut off the lid of the cigar box.
3. Place the face of the cigar box on corrugated cardboard and trace the shape with a pencil. Cut out the cardboard and cover it with colored paper.
4. Fit the cardboard in the box at a slant facing upward to see if it fits (figure 6-13A).
5. Take the cardboard out and arrange the vials on it in the sequence to be displayed. Fasten the vials to the cardboard by punching one small hole on each side of a vial and running a piece of flexible wire through one hole, around the vial and through the other, or twisting it in back of the cardboard (figure 6-13B).

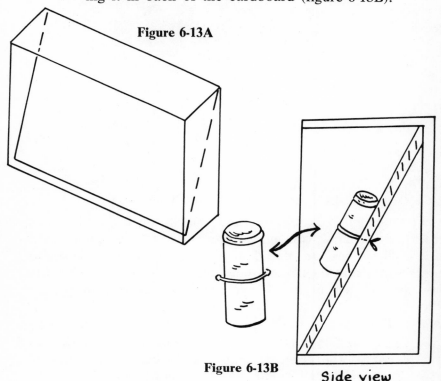

Figure 6-13A

Figure 6-13B   Side view

6. Place appropriate labels to identify the insect and the stage of its life cycle (figure 6-14).

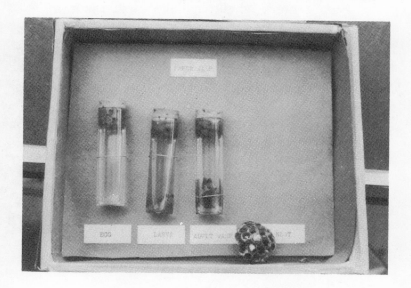

Figure 6-14

MAKING AN INSECT-REARING CAGE

Many students while studying and collecting insects want to watch the life cycle in the classroom. Encourage them to donate circular cartons to build rearing cages.

Materials:

Window screen mesh, circular carton with a lid such as for oats cereal or ice cream, and masking tape.

Procedure:

1. Cut out a hole about 5½ inches high and 6 inches wide in the side of the circular carton.
2. Cut out a piece of window screen mesh 6 inches high and 7 inches wide.

3. Curve the mesh and place it behind the opening in the carton. Tape all around the edge of the screen on the inner surface of the carton (figure 6-15). Specimens can be observed through the mesh window.

Figure 6-15

PREPARING PERMANENT MICROSCOPE SLIDES OF INSECT PARTS

Because insects dehydrate or dry quickly, students may make permanent microscope slides of the parts of insects such as butterfly scales, mouth parts, wings or any small part of an insect.

Materials:

Canada balsam, microscope glass slides, glass cover slips, gummed labels, toothpicks and a dried insect.

Procedure:

1. Submerge the end of a toothpick in Canada balsam and place three or four drops of it in the center of a microscope glass slide.

2. Place the insect part on top of the balsam. Take a clean toothpick and submerge and center the part. Use the toothpick to remove or break any air bubbles.
3. Carefully place a glass slide over the balsam. Gently apply pressure on the glass slide so the balsam will spread toward the edge. Store the slides to dry in a horizontal position. They will take about a week to dry, depending upon the humidity. When they are dried, students may glue small labels at the edge of the slides for identification.

CULTURING MEALWORMS

One way to study the life cycle of an insect in the classroom is to culture mealworms. The mealworm is actually a larva which develops into a grain beetle, *Tenebrio molitor*. Therefore the grain beetle goes through a complete metamorphosis: egg, larva (mealworm), pupa and beetle. You will get a complete life cycle in from 4 to 6 months. The larval stage can be slowed by temporarily storing in a refrigerator. The adult is considered a pest and often invades granaries (figure 6-16). Mealworms can generally be purchased from your local pet shop or from a biological supply house.

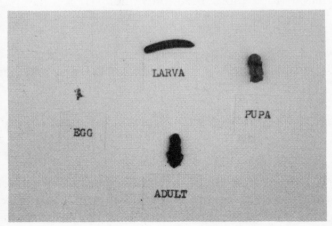

Figure 6-16

Materials:

Jar or can, breakfast bran or cereal flakes, carrot, apple or celery, and a cheese cloth or window screen mesh.

Procedure:

1. Half fill a jar or can with breakfast bran or cereal flakes.
2. Place pieces of carrot or apple near the surface. This will supply the moisture needed to keep them alive.
3. Now add your culture of mealworms. They will burrow down into the cereal.
4. Place a piece of cheese cloth or screen mesh over the top of the jar to protect them from other insects.

Fresh carrots and cereal should be added as the old becomes pulverized. About once a month remove any debris including bodies of dead beetles. Mealworms can be used in the classroom as food for birds, frogs. lizards, salamanders and snakes.

INVESTIGATING MEALWORMS

Mealworms can be purchased from a local pet shop and can easily be cultured in the classroom. Students may be given mealworm larvae to explore their movements.

Materials:

Mealworm, ½ of a culture dish (petri) without the cover, microscope glass slice, 3 × 5 inch index card and hand lens.

Investigation No. 1 — How does a mealworm explore a culture dish?

Procedure:

Trace a culture dish on a sheet of paper. Place a mealworm in the center of the culture dish and leave it free to explore. As the mealworm moves in the dish, the students should trace its course with dotted lines on the sheet of paper. After observing the mealworms, see if the students can make a general statement about the exploration of the mealworm.

Investigation No. 2 — How does a mealworm react to light and darkness?

Procedure:

Trace a culture dish on a sheet of paper. Cover the left half of the dish with a 3 × 5 inch card. Place a mealworm in the center of the dish (figure 6-17). Have the students trace its course in dotted lines on the sheet and make a general statement on how the worm reacts to the cover.

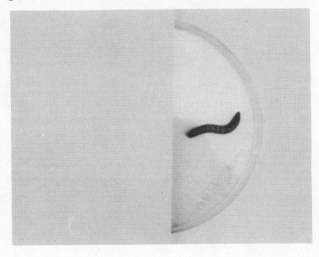

Figure 6-17

Investigation No. 3 — How does a mealworm react to an obstacle?

Procedure:

Trace a culture dish on a sheet of paper. Place a microscope slide in the center of the dish. Place a mealworm to the right of the glass slide (figure 6-18). Have the students follow the course of the mealworm on their papers to see if it goes around or over the glass obstacle. Have them make a summary statement about its behavior.

Investigation No. 4 — Observation of the mealworm body structure.

Figure 6-18

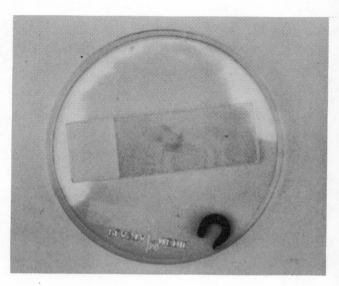

Procedure:

Have the students place a mealworm in a culture dish and observe its body structure with a hand lens. Ask the students if it has any eyes, antennae, palps (tasters), and ask the number of legs and body segments. Have the students make a drawing of the mealworm and label any structures observed.

Encourage students to devise their own investigations with the mealworm. They may wish to study, observe and make displays of its life cycle.

OBSERVATION OF AN INSECT

This is a lesson for students on the observation of an insect. The teacher may prepare a ditto sheet of the points to be observed. The students will have to supply their own insects for this activity.

Materials:

One insect, hand lens, forceps and a sheet of questions.

Procedure:

1. What color is the body of your insect?
2. Feel the body with your fingers. Does it seem to have a hard covering?
3. How many large body regions does it have?
4. How many wings does it have? To what part of the body are the wings attached?
5. How many large eyes do you find? Examine the eyes with a hand lens. Are they made up of one or many parts?
6. How many antennae (feelers) does the insect have?
   a. Where are they located?
   b. Are they made up of knobs or one straight piece?
7. How many legs does it have?
   a. To which part of the body are the legs attached?
   b. Are all the legs the same length?
8. Can you see any mouth parts under the head such as tasters (palps)?
9. Look at the back end of the body (abdomen) and see if you can find any tiny breathing pores (spiracles) (figure 6-19).

Figure 6-19

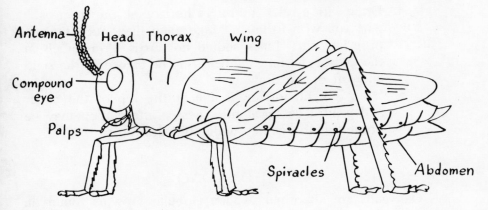

MAKING A DRAWING OF AN INSECT

Students should now learn some standard procedures on how to make a drawing and label its parts. An insect or its parts can serve this purpose. The drawing should represent the observations made by the student and should be made with clear, clean-cut lines with correct labeling. Drawing should be done in pencil.

Materials:

> One insect or parts of one, hand lens, pencil, ruler and underlined paper.

Procedure:

1. If you are making a drawing of one insect, it should be in the center of the paper.
2. Estimate the size of your drawing. It should be in proportion to the size of the paper. Most drawings at this grade level are either too large or too small. Make the drawing large enough to see what it represents. Leave enough space for labels.
3. Sketch in the drawing with a pencil. When the drawing is completed erase all extra lines. The drawing should be neat with clear, clean lines. Shading of the drawing should be discouraged.
4. Draw with a ruler a straight horizontal line from each part you wish to label. Make each line touch the appropriate part. Lines should not cross. Print labels at the end of the lines. Labels may be placed to the right and left of the drawing.
5. Draw a light horizontal line at the top of the paper and print a centered title. Lesser information may be printed at the center under the drawings.

THE OAK GALL

Oak galls are abnormal swollen globular growths found on the bark or stems of oak tree leaves. The average full grown gall is approximately three inches long and 2½ inches wide. The interior is largely a solid spongy mass.

It is believed that the growth of a gall in tree tissue is caused by the larvae of the oak gall wasp *(Cynipidae)*. The wasp is a parasite upon the oak tree. The gall is rich in protein and gives the insect food, shelter and protection from birds. The insect spends most of its life enclosed in the gall. From a larva it develops into a pupa and then an adult. The adult wasp emerges by burrowing to the outside (figure 6-20).

Figure 6-20

Observations of the oak gall — the "Thing"

If oak trees are in your locality, spend some time looking for and collecting oak galls. These galls will be used as a lesson in observation. This activity is carried out individually or in groups of two, depending upon how many galls you collect.

Materials:

One oak gall, 1 sheet of unlined paper, 1 sheet of lined paper and hand lens.

Procedure:

1. Give out the galls to the students. Do not tell them what they are.
2. Ask the students to place the title "The Thing" on their papers, since they do not know what it is. Have them make a drawing of the gall and count the number of holes in it.
3. On the lined paper have them set up written questions about the gall. Some of the questions asked might be: Is it hollow? What is it made of? Is it a seed?

Collecting the life cycle

Materials:

One oak gall, forceps, petri dish (culture dish) and hand lens.

Procedure:

1. Hand out to each group of students a hand lens, forcepts, petri dish, and an oak gall split in half. The teacher should split the galls in half with a hammer and chisel.
2. Have them look inside the gall and take out any of the stages that may be found, such as the larva, pupa or adult, and place them ·in a petri dish to be examined with a hand lens.

For students to complete the investigation, have them make a display of the life cycle by placing each stage in pill bottles or vials with alcohol or by making large display charts (figure 6-21).

Figure 6-21

The chemistry of the oak gall

Testing the gall solution for acid or base

A gall solution will have to be manufactured before it can be tested for acid or base.

Materials:

One gall, course sandpaper or file, filter paper, water on heat source, funnel, and blue and pink litmus paper.

Procedure:

1. Pulverize ½ of a gall with a file or sandpaper.
2. Place the gall powder into a beaker of boiling water. Let the water boil several minutes and filter the solution through a paper filter (or paper towel) in a funnel into a clean beaker. The solution should be similar in color to tea. Let the solution cool.
3. Have the students test the solution with blue and pink litmus paper. If the blue litmus paper turns pink, the solution is acid. If the pink litmus paper turns blue, the solution is base. They will see that the gall solution will turn blue, indicating the solution is acid. The gall solution contains tannic acid.

Making a dye and ink with tannic acid

This will give students an opportunity to make some blue dye and ink from the gall solution containing tannic acid when it is mixed with an iron compound.

Materials:

Gall solution (tannic acid), iron compound (ferric ammonium sulfate), 1 square inch of white cloth and a paper drinking cup.

Procedure:

1. Pour some of the tannic acid solution into a paper cup. Place 1 square inch of white cloth into the tannic acid solution, and take it out and let dry overnight.
2. Make a small amount of ferric ammonium sulfate solution by mixing about 1 gram of it in about half a test tube of water.
3. Pour a small amount of the solution onto the cloth dried the night before. The cloth will change to a blue color indicating that it was dyed. Let the dyed cloth dry so it can be observed the next day.
4. Blue ink may be made by mixing about five drops at a time of tannic acid solution and ferric ammonium sulfate solution in a paper drinking cup. The solution

will instantly turn blue. If fountain pens are available to the students, have them write a sentence with their homemade ink.

Students may want to test other materials as coffee and tea for tannic acid. An iron compound, ferric ammonium sulfate, can be used as an indicator. When it turns a solution blue, this indicates the presence of tannic acid.

Tannic acid is used in making inks and dyes, medicines and insecticides, and tanning leather. Students may make some kind of displays or charts illustrating an activity about the gall (figure 6-22).

Figure 6-22

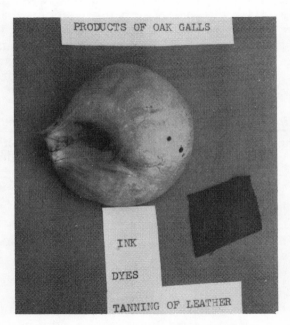

PRODUCTS OF OAK GALLS

INK

DYES

TANNING OF LEATHER

DEVELOPING AN INSECT ORDER WHEEL

Materials:

Two sheets of unlined paper, 1 brad fastener, 1 ruler, scissors, and a drawing compass.

Procedure:

1. Draw a circle with a drawing compass about 6 inches in diameter on a sheet of unlined paper. Then cut out the circle.

2. Take a ruler and draw two rectangles 1 1/8 inches long and ½ inch wide alongside each other left of center circle. Draw a third right of center circle.
3. With scissors cut out the three rectangles. These will now look like three windows.
4. Above the first window to the left, print the label "Order". Below it print the label "Means". Above the second window print "No. of Wings". On top of the third window print "Development". Print a title on the upper half of the circle such as "Insect Order Wheel".
5. Take a paper brad fastener and place it through the center of the circle. Then pass the brad through the center of an unlined sheet of paper and fasten. The circle serves as a rotating wheel.
6. The students may develop the insect order wheel by placing the name of an insect such as termite on the left side of the paper outside the wheel. The order of the termite, *Isoptera,* may be printed in the first window with its meaning "equal-wing" underneath it. In the second window can be placed the number of wings, 0-4. In the window to the right can be placed its development, incomplete (figure 6-23).

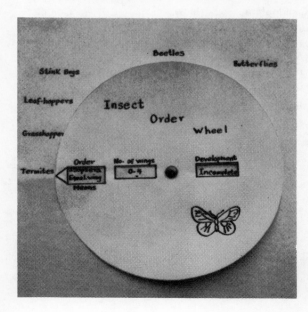

Figure 6-23

Students may continue turning the wheel and placing names of insects outside and vital information about them inside the wheel. Then students may use the wheel as a reference.

The following information may be useful to the teacher in helping students to develop the Insect Order Wheel.

ORDER OF INSECTS

Insects	*Order*	*Means*	*No. of Wings*	*Development (Metamorphosis)*
Termites	Isoptera	Equal-wing	0-4	Incomplete
Dragonflies	Odonata	Tooth-jawed	4	Incomplete
Damselflies	Odonata	Tooth-jawed	4	Incomplete
Grasshoppers	Orthoptera	Straight-wing	0-4	Incomplete
Crickets	Orthoptera	Straight-wing	0-4	Incomplete
Roaches	Orthoptera	Straight-wing	0-4	Incomplete
Leaf-hoppers	Homoptera	Same-wing	0-4	Incomplete
Aphids	Homoptera	Same-wing	0-4	Incomplete
Stink bugs	Hemiptera	Half-wing	4	Incomplete
Water Striders	Hemiptera	Half-wing	4	Incomplete
Beetles	Coleoptera	Sheath-wing	4	Complete
Butterflies	Lepidoptera	Scale-wing	4	Complete
Moths	Lepidoptera	Scale-wing	4	Complete
Bees	Hymenoptera	Membrane-wing	0-4	Complete
Wasps	Hymenoptera	Membrane-wing	0-4	Complete
Ants	Hymenoptera	Membrane-wing	0-4	Complete
Flies	Diptera	Two-wing	0-4	Complete
Mosquitoes	Diptera	Two-wing	0-4	Complete
Fleas	Siphonaptera	Siphon-mouth	0	Complete

APPENDIX

Science Supply Houses

The following is a list of science supply houses which can be useful to the teacher when ordering science supplies:

CENTRAL SCIENTIFIC COMPANY (CENCO)

160 Washington Street
Somerville, Massachusetts 02143

237 Sheffield Street
Mountainside, New Jersey 07092

2600 S. Kostner Avenue
Chicago, Illinois 60623

6446 Telegraph Road
Los Angeles, California 90022

1040 Martin Avenue
Santa Clara, California 95052

6901 E. Twelfth Street
Tulsa, Oklahoma 74115

15 N. 40th Place
Phoenix, Arizona 85034

6610 Stillwell Street
Houston, Texas 77017

2128 Seventh Avenue
Birmingham, Alabama 35233

CURTIN & COMPANY, W. H.

P.O. Box 5304 (75222)
1103-07 Slocum Street
Dallas, Texas

P.O. Box 386
470 Valley Drive
Brisbane, California 94005

P.O. Box 2447
15215 Marquardt Avenue
Santa Fe Springs, California 90670

750 Adams Avenue (38105)
Memphis, Tennessee

P.O. Box 747 (74101)
6550 E. 42nd Street
Tulsa, Oklahoma

P.O. Box 2122 (30301)
1782 Marietta Blvd., N.W.
Atlanta, Georgia

FISHER SCIENTIFIC COMPANY

The Stansi Scientific Division
1231 North Honore Street
Chicago, Illinois 60622

GENERAL BIOLOGICAL, INC. (TURTOX)

8200 South Hoyne Avenue
Chicago, Illinois 60620

342 Western Avenue
Boston, Massachusetts 02135

1945 Hoover Court
Birmingham, Alabama 35226

LA PINE SCIENTIFIC COMPANY

920 Parker Street
Berkeley, California

375 Chestnut Street
Norwood, New Jersey 07648

6001 S. Knox Avenue
Chicago, Illinois 60629

NASCO (SCIENCE SUPPLIES)

Fort Atkinson
Wisconsin 53538

P.O. Box 3837
Modesto, California 95352

SARGENT-WELCH SCIENTIFIC COMPANY

7300 N. Linder Avenue
Skokie, Illinois 60076

609 W. 51st Street
New York, N.Y. 10019

15233 Ventura Blvd.
Sherman Oaks, California 91403

VAN WATERS & ROGERS SCIENTIFIC SUPPLIES CO.

3950 N. W. Yeon Avenue
Portland, Oregon 97210

600 S. Spokane Street
Seattle, Washington 98134

WARDS NATURAL SCIENCE ESTABLISHMENT, INC.

P.O. Box 1749
Monterey, California 93940

P.O. Box 1712
Rochester, New York 14603

BIBLIOGRAPHY

BARRETT, RAYMOND E., *Build-It-Yourself Science Laboratory,* Garden City, New York: Doubleday and Co. Inc., 1963.

BEELER, NELSON F., and BRANLEY, FRANKLYN M., *Experiments With Electricity,* New York: Thomas Y. Crowell Co., 1949.

BLOUGH, GLENN O., and SCHWARTZ, JULIUS, *Elementary School Science,* New York: Holt, Rinehart and Winston, Inc., 1964.

BORROR, DONALD J., and DELONG, DWIGHT M., *An Introduction to the Study of Insects,* New York: Rinehart and Co., 1957.

BRANDWEIN, P., JOSEPH, A., and MARHOLT, E., *A Sourcebook for the Biological Sciences,* New York: Harcourt, Brace and Co., 1966.

BRANDWEIN, P., JOSEPH, A., MARHOLT, E., POLLACK, H., and CASTKA, J., *A Sourcebook for the Physical Sciences,* New York: Harcourt, Brace and Co., 1961.

BRANLEY, FRANKLYN M., *Experiments in Sky Watching,* New York: Thomas Y. Crowell Co., 1959.

BRENT, ROBERT, *The Golden Book of Chemistry Experiments,* New York: Golden Press, Inc., 1960.

BURGDORF, OTTO P., *The Question and Answer Book of the Human Biology,* New York: Capital Publishing Co., Inc., 1962.

COOPER, ELIZABETH K., *Discovering Chemistry,* New York: Harcourt, Brace and Co., 1958.

EDUCATIONAL SERVICES INCORPORATED, *Behavior of Meal-worms — Teachers Guide,* Boston: Houghton Mifflin Co., 1964.

FEIFER, NATHAN, *Let's Explore Chemistry,* New York: Sentinel Books Publishers, Inc., 1959.

FREEMAN, IRA MAE, *Fun With Chemistry,* New York: Random House, 1944.

GENERAL BIOLOGICAL SUPPLY HOUSE, *Turtox Service Leaflets,* Chicago: General Biological Supply House.

GRAY, ALICE, *The Questions and Answer Book of Insects,* New York: Capital Publishing Co., Inc., 1963.

HANAUER, ETHEL, *Biology for Children,* New York: Sterling Publishing Co., Inc., 1962.

HILLCOURT, WILLIAM, *Field Book of Nature Activities,* New York: G. P. Putnam's Sons, 1950.

HONE, E., JOSEPH, A., and VICTOR, E., *A Sourcebook for Elementary Science,* New York: Harcourt, Brace and Co., 1962.

KLOTS, ALEXANDER B. and ELSIE B., *1001 Answers to Questions About Insects,* New York: Grosset and Dunlap Publishers, 1961.

LIBERTY, GENE, *The How and Why Wonder Book of Time,* New York: Wonder Books, Inc., 1963.

MAY, JULIAN, *There's Adventure in Rockets,* Chicago: Popular Mechanics Press, 1958.

MORGON, ALFRED, *First Chemistry Book for Boys and Girls,* New York: Charles Scribner's Sons, 1950.

NATIONAL SCIENCE TEACHER'S ASSOCIATION, *Ideas for Teaching Science in the Junior High School,* Washington, D.C.: N.S.T.A. (1201 Sixteenth St., N. W.), 1963.

NELSON, PEARL A., *Elementary School Science Activities,* Englewood Cliffs, N. J.: Prentice-Hall, Inc., 1968.

RUCHLIS, HY, *Orbit: A Picture Story of Forces and Motion,* New York: Harper and Brothers Publishers, 1958.

SCHEINFELD, AMRAM, *The New You and Heredity,* New York: J. B. Lippincott Co., 1950.

SCOTT, JOHN M., *The Everyday Living Approach to Teaching Elementary Science,* West Nyack, New York: Parker Publishing Company, Inc., 1970.

SOOTIN, HARRY, *Experiments in Magnetism and Electricity,* New York: Franklin Watts, Inc., 1962.

Experiments With Electric Current, New York: W. W. Norton and Co., Inc., 1969.

STEINBERG, WILLIAM B., and FORD, WALTER B., *Electricity and Electronics — Basic,* Chicago: American Technical Society, 1961.

SWAIN, RALPH B., *The Insect Guide,* Garden City, New York: Doubleday and Co., Inc. 1948.

TANNENBAUM, BEULAH, and STILLMAN, MYRA, *Understanding Time — The Science of Clocks and Calendars,* New York: Whittlesey House, McGraw-Hill Book Co., Inc., 1958.

UNESCO, *700 Science Experiments for Everyone,* Garden City, New York: Doubleday and Co., 1965.

VICTOR, EDWARD, *Exploring and Understanding Magnets and Electromagnets,* Chicago: Benefic Press, 1967.

VIORST, JUDITH, *150 Science Experiments Step-by-Step,* New York: Bantam Books, Inc., 1967.

VRANA, RALPH S., *Junior High School Science Activities,* Englewood Cliffs, N. J.: Prentice-Hall, Inc., 1969.

WITHERSPOON, JAMES D., and REBECCA, H., *The Living Laboratory,* Garden City, N. Y.: Doubleday and Co., 1960.

WYLER, ROSE, *The First Book of Science Experiments,* New York: Franklin Watts, Inc., 1952.

INDEX

Y